dakini books

Published in 2006
by dakini books NP
T 020 7830 9692
F 020 7830 9693
www.dakinibooks.com

Reprinted in 2006

Published under licence in the UK
by Guardian Books and A & C Black Publishers Ltd
38 Soho Square, London W1D 3HB
www.acblack.com

© dakini books NP

ISBN-10: 0 7136 8205 1
ISBN-13: 978 0 7136 8205 2

A CIP catalogue record for this book is available
from the British Library

Publisher: Lucky Dissanayake
Author: Paul Brown
Advisors: James Cameron, Gerard Wynn,
Phil Wolski, Hélène Connor
Editor: Paul Olive
Design: Paul West, Paula Benson, Andy Harvey – Form®
Research: Hayley Furminger, Emily Chew
Reproduction: Richard Deal – Wyndeham Prepress

Printed on GardaPat 13 KIARA 135 gsm
by Cartiere del Garda
Cartiere del Garda is committed to
sustainable development
ISO 14001 Certified and EMAS registered

Mixed Sources
Product group from well-managed
forests and other controlled sources
www.fsc.org Cert no. SW-COC-1221
© 1996 Forest Stewardship Council

Printed and Bound in Europe

10 9 8 7 6 5 4 3 2

The carbon emitted at the publishers during the
production of this book has been offset

GLOBAL WARNING

The Last Chance for Change

PAUL BROWN

 theguardian

Contents

Foreword: His Excellency Maumoon Abdul Gayoom, President of the Republic of Maldives.

For countries like the Maldives, global warming could very well become a matter of life and death.

Indeed, as a frontline nation our low lying islands will be amongst the first to suffer the consequences of sea level rise due to global warming. Sea level rise, if not stopped in time, could totally submerge the Maldives and other small island states. The 2004 tsunami disaster highlighted this very fragility, as for a few harrowing minutes, almost our entire archipelago went under water. The tsunami claimed more than lives and livelihoods. It gave face to a concern that we have been addressing for a number of years — that as long as we kept ignoring the real danger of global warming the Maldives, and indeed, all low lying areas, lived on borrowed time.

The author of this book, Paul Brown, has used his vast experience as an environment correspondent for the Guardian to present these compelling facts about global warming and climate change in a form that is comprehensible to the layman. His presentation of the complex web of inter-linkages between science, business and politics in the climate change debate is thought-provoking.

This book is a clarion call to all those who ignore years of extensive scientific research and proof that global warming is real and an impending major threat to the whole world. However, the message that Paul Brown gives in this book is that we can still turn back from this path of doom, and that there is still a chance to change. Indeed, the book presents a number of measures that can be taken at an individual, community, regional and international level to reverse global warming.

The challenge that we face is, doubtless, daunting. But human beings, as the most resilient and intelligent inhabitants of planet Earth, have the knowledge, the wisdom and the resources to overcome even this challenge. What is needed is the will to act.

For centuries we have lived on the rich bounty of the earth. It is time that we gave something back. We owe this to Mother Earth.

Our Changing World

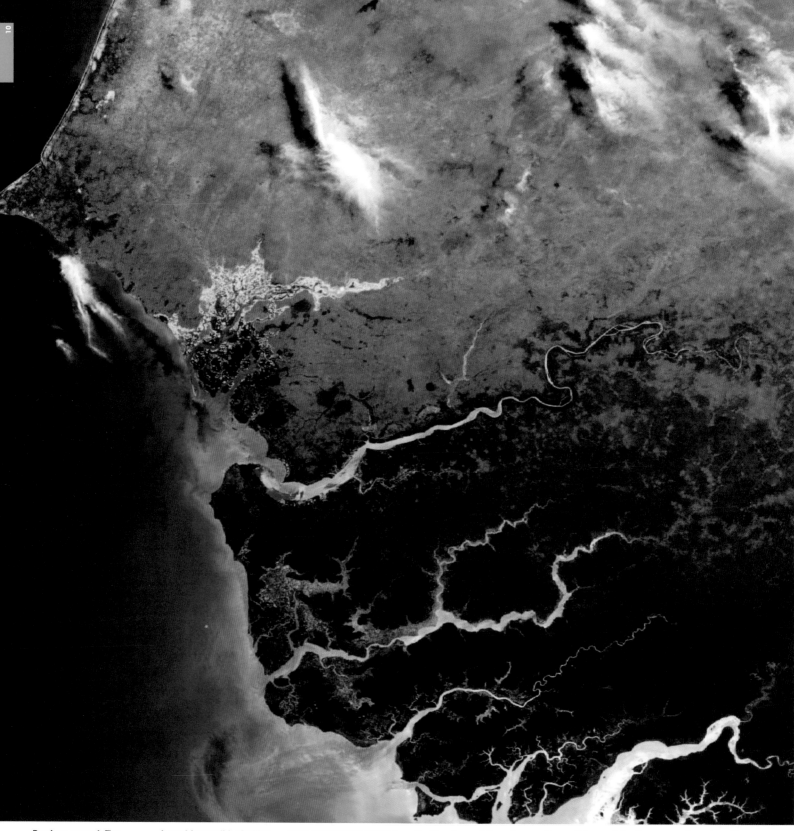

Previous spread: The permanent bright lights of Earth seen from space. This remarkable image constructed from satellite photographs shows the most urbanised areas of the planet, but not necessarily the most populated. Compare China and India with western Europe. The outlines of a world map have been superimposed for guidance; it is startling how many centres of population are on the coast, and are vulnerable to sea level rise. It is clear that the United States, Europe and Japan are using large amounts of electricity for lighting cities at night. It is a sobering thought that most of the light comes from the burning of fossil fuels. Two lines on the map are of note, the bright squiggle in eastern North Africa shows the concentration of population along the Nile, and across Russia, the light marks the route of the Trans-Siberia railway from Moscow to Vladivostok. Dark areas are the great deserts in Africa, Arabia, Australia, Mongolia, China and the United States, and the mountainous areas of the Himalayas. Although Iceland can be seen, Greenland and Antarctica remain dark.

Above: It is surprising how much of what is happening to natural systems on Earth can be monitored from space. This picture covers the transition zone between the desert and savannah in the north of Africa and the tropical vegetation further south. Clearly seen is the discharge into the sea of sediment via the river system along the west coast. This is the result of soil erosion, partly caused by overgrazing and the loss of vegetation. A series of images over time can keep track of changes and pinpoint where intensification of land use is causing problems and action is needed on the ground to halt the erosion.

Above: The Sahara desert will spread north and leap the Mediterranean, according to scientists studying climate change. This may sound fantastic but, as this picture taken on November 14th, 2004 shows, vast clouds of dust can be carried out over the sea from North Africa. This image shows Libya in the centre and Tunisia top left. Much of the dust fell in Europe over several days as a large weather system whipped up gale force winds which affected Algeria, Italy and as far north as Albania. Arid conditions in southern Spain, Portugal, Italy and Greece were already causing concern before the prolonged drought and heat wave in 2005. All four countries are vulnerable to climate change, which will in this region bring extra heat and prolonged droughts. Because of the deteriorating conditions these southern European states have all joined the UN's Desertification Convention.

2002

2003

2004

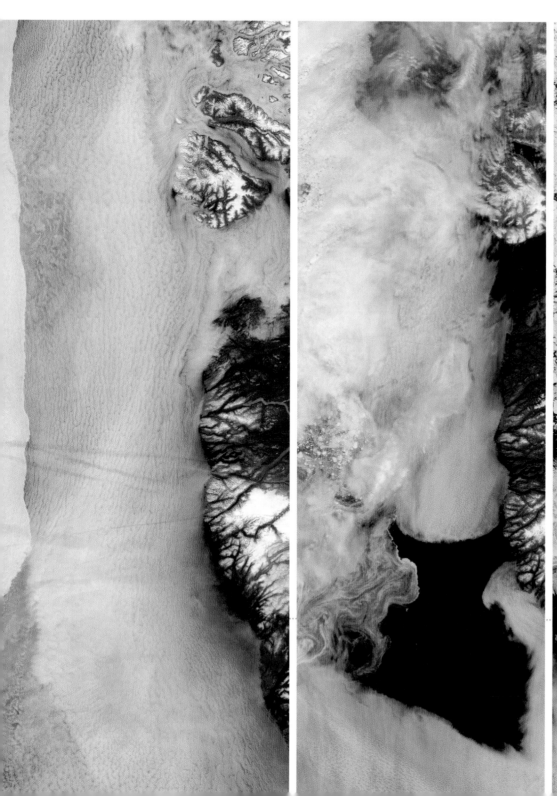

Climate change is the most important issue of the 21st century. The global economy, and civilisation itself, may collapse unless greenhouse gas emissions are controlled. Already global warming touches every part of the planet and people everywhere are affected in their daily lives.

Changing weather patterns and the need to adapt to new conditions will dominate policy as the earth's temperature and sea levels continue to rise.

Scientists believe that time is already very short — there may be as little as 15 years to prevent irreversible climate change. Yet politicians, who have been made aware of the dangers, continue to act as though it were some far off threat. As carbon dioxide builds up in the atmosphere at an ever faster rate, leaders of nations talk about the need for more talks to discuss how to solve the problem.

This book looks at the key issues. The science, the politics, what is happening in today's threatened world and what (if anything) can be done. It is extraordinary how many people still do not understand the danger the planet is in. The information is available to anyone who cares to investigate, and so there are some who believe that most of the developed world is in denial. They pick as an example the people who drive fashionable four-wheel-drive cars and take cheap flights to holiday homes. They are the same people who at the same time make incredible efforts to give their children the best education and start in life, on the assumption that life will be much the same in 50 years. It won't be. In Europe governments are worrying about pensions because of an ageing population. They are exhorting 30-year-olds to start planning for their pensions. These young people, at least those who know about climate change and the economic havoc it will cause to the "safest" of our financial institutions, just laugh. They are almost certainly right to do so. For those who are in any doubt that climate change is real and happening now, this book explains the issues. It explores other related environmental problems, which are made worse

by climate change. Reducing poverty, improving water resources and sanitation, cutting air pollution, controlling tropical diseases and saving species from extinction are all important. Add to that the need to increase food production to feed an ever rising population, on what will soon be significantly smaller land masses, and you will see the difficulty.

The pictures show the changes that are already obvious. Photographs from space document disappearing ice, forests and the dust storms from enlarging deserts. To bring this down to the particular there are pictures of creatures in the wild which are already disappearing and may be extinct in the lifetime of our children — or at least kept alive only in zoos because their natural habitat has disappeared. Polar bears are a good example.

But while it may be sad to lose these, and numerous other less majestic creatures, their fate is only an illustration of what may happen to much of the human race. Sea level rise is going to wipe out the homes of millions of people. Already islands in the Indian and Pacific oceans are being evacuated because of rising waters. People are having to abandon homes their forebears have lived in for thousands of years. They are being made uninhabitable by saline intrusion into water supplies and high tides overtopping their islands. Other island countries, like the Maldives, have begun fortifying some islands as "safe havens" against the sea. How long will they last?

That is only the beginning. The small island states, the idyllic palm-fringed islands of the holiday brochures, are home to a few million people. But in Bangladesh 15 million live less than a metre above sea level, in India there are another 8 million. These are the people with

Left: The west coast of Greenland showing the speed of change between 2002 and 2004 all taken at the same time in June from an American satellite. Temperatures in Greenland and the Arctic generally have increased far faster than in the tropics, leading to more rapid ice melt than scientists had predicted and accelerating the rise in sea level. Scientists recently documented that the seasonal thaw along the margins of the Greenland ice sheet has been starting much earlier in the year and impacting a larger area than in the past century. As can be seen, sea ice in the region has been declining at the same time. More pronounced than the differences in completely snow-free areas are the increases in the total area caught in the act of melting — a translucent area sandwiched between the bare ground near the coast and the solid white of the interior of the island.

little money and no protection who will have to migrate to higher land. In an already crowded continent it is hard to know where they can go.

But sea level rise does not just affect the poor. Many of the world's largest and richest cities are at sea level. Dozens of cities will suffer total or partial inundation if they do not raise defences against the sea. London and Rotterdam already have barriers that can be swung into place when high tides or storm surges threaten to overwhelm their defences. St Petersburg and Venice also realised some time ago that they are doomed without massive engineering works to keep out the sea. Fortunately for the people who live in these cities they are on estuaries or lagoons where, with modern technology, barriers can be built across the entrances with gates to be closed when dangers threaten. Others cities, including New York, and Chittagong in Bangladesh, are much more exposed.

Sea level rise is just one of the perils the planet faces. Climate change makes a drastic difference to the food supply. The poorest continent, Africa, is already suffering from food shortages made far worse by a series of droughts, believed to be at least partly caused by climate change. In all parts of the world farmers are having to adapt to new temperature and rainfall patterns, and grow different crops to suit the changed conditions. Mostly this leads to lower productivity and sometimes forces them to abandon the land altogether. Already more refugees flee because of environmental factors than from war, according to the United Nations. In 2005 the news about climate change was bad. Global temperatures equalled those of the warmest year on record, 1998. It meant that the unprecedented series of warm years from the last decade continues. New data confirmed

Left: Flood waters reflecting the grand façade of Belgrade's main train station on April 17th, 2006 when the Danube broke through flood defences. Thousands of people were forced from their homes in Serbia, Romania and Bulgaria. The combination of rapid snow melt in central European mountains and heavy rain caused this flood, one of a series in the last five years when major rivers in Europe have burst their banks. Increasingly large areas of Europe can no longer obtain flood insurance as the number of these incidents has grown over the last 10 years.

Top: Extreme weather events, as scientists call floods, windstorms and droughts, are increasingly common all over the world. Here rescue personnel search for people trapped in cars on a flooded street in Buenos Aires during torrential overnight rain, January 31st, 2004.

Above: Flash floods have always been a feature of tropical regions, made worse by tree-cutting and urban development. Vegetation soaks up rain while hard surfaces turn roads into rivers. Warming oceans increase the violence of tropical storms. Here in October 2005 a Haitian man holds a woman in danger of being swept away during a storm in Port-au-Prince. The tropical storm Alpha brought heavy rain, flash floods and mudslides to Haiti and the Dominican Republic.

some of the worst fears of scientists about existing warming trends and added some new and unexpected dangers. It is not just the addition of carbon dioxide from burning fossil fuels that is making the situation worse, the extra heat is believed to be releasing carbon locked in soil, and methane in permafrost. Parts of the Arctic were warmer than ever previously recorded, hastening the disappearance of sea ice and glaciers. Vast areas of permafrost, with millions of tonnes of trapped greenhouse gases, began to melt in Siberia. In 2005-06 Svalbard (otherwise known as Spitsbergen), the group of islands well to the north of Norway, had its warmest winter since records began. At the opposite pole, the West Antarctic ice shelves, thought to be stable, began to slide into the sea.

Scientists also discovered that the extra carbon dioxide in the atmosphere was making the oceans more acid. After millions of years of creatures adapting to existing conditions (for example corals and shellfish relying on the alkaline sea water to extract the raw material for their shells), what will the effect be? The whole balance of the oceans may change. It has also become a lot clearer that sea temperatures are rising. Fishermen, who for centuries have hauled the same kinds of fish out of the same stretches of water, are finding that they have died out or migrated to find cooler places and been replaced by new exotic varieties. There are new fears that the ocean currents which transport heat round the globe are changing — and that the Gulf Stream might slow down or shut off altogether, plunging parts of Europe into a much colder climate.

Partly because of increasing sea temperatures 2005 was another year of extreme weather events. Over the last two decades these have steadily become more common the world over. Floods and droughts, which make headlines every year, were eclipsed because of the largest and most severe series of hurricanes ever recorded, made worse by the higher sea temperatures in the southern North Atlantic and in the Gulf of Mexico. Despite being the richest country in the world, with the best technology for predicting and studying hurricanes, the United States could not cope with the storm named Katrina, which killed more than 1,300 people and devastated New Orleans. This is a city below sea level, prone to more hurricanes, aware of its vulnerability, yet it still failed to prepare for its fate. Despite this traumatic, tragic and expensive lesson the US government is intent on spending billions of dollars rebuilding New Orleans in the same place. This is the sort of policy decision that leads some to believe that ours is a culture and a civilisation in denial. This would not be the first. Civilisations have collapsed and disappeared in the past because of over-consumption. In most cases it has been overuse or misuse of water supply. In others over-hunting and over-fishing wiped out the available food sources. In modern times Haiti is a classic example. The country of 8 million people was once largely covered in forests but only 2% remain because most were cut for firewood. The loss of forests meant the soil was washed away. The country is now suffering ecological and economic collapse and only international aid is keeping it alive.

In every case the warning signs of impending disaster must have been there but for some reason the political leaders failed to act in time. In the same way our civilisation is steadily removing all its life support systems at once, on a global scale, and at the same time changing the climate. It is not just events like those in New Orleans that lead to the conclusion that

"Climate change is no longer an abstract subject for research or a highly contested issue for political debate. Climate change is a reality – one of the most serious threats facing humanity today."

Richard Kinley, officer-in-charge of the United Nations climate change secretariat, speaking at the Montreal climate change conference, December 8th, 2005.

Top: Mongolian villagers migrate across their vast country in search of ever decreasing grassland regions to feed their stock and escape from the ever spreading deserts. Further south Chinese scientists report that there are "desert refugees" in three provinces: Inner Mongolia, Ningxia and Gansu. The Asian Development Bank's preliminary assessment of Gansu province has identified 4,000 villages that will probably have to be abandoned as desertification intensifies.

Above: Thousands of internally displaced Kenyans wait in line at a water point on the road to the northern town of Wajir, 500 km (300 miles) from the capital Nairobi, January 12th, 2006. Kenya waived import duty on relief food to feed millions of people facing famine in the country's worst drought for years. The drought affected large parts of eastern and southern Africa, where the rains have become less reliable, as climate scientists predicted would happen as the climate warms. Partly as a result of the increasing difficulty of survival in the countryside people have moved into shanty towns on the edges of cities in the hope of making a living.

Above: A Chinese boy hugs his dog in front of his family home, which will be demolished after the foundations were damaged by the water level rise along the Yangtze River in Zigui, central China. The water was rising with the filling of the mammoth Three Gorges dam reservoir. The area is higher than the water level of the reservoir and was supposed to be safe. China moved a total of 1.3 million people from areas affected by the controversial dam project which is designed to provide hydro-electric power and control the flow of the river. The Chinese character reads "Demolish".

our political leaders are unable to face up to the crisis in civilisation they are helping to create. There is a point that needs to be made here about the responsibility of all of us, including scientists. Several times in researching this book I have been told by some scientists that they cannot be sure that hurricane Katrina or the melting of the ice caps is related to man-made climate change. It is what the computer models predict would happen, and the sort of events they would expect, but they cannot as yet be certain that any one event is caused by the man-made greenhouse effect and is not just a natural phenomenon. This is of course true, but put together the evidence of all the remarkable weather events that occurred round the world in 2005, 2004 and the decade before and it seems to me overwhelmingly evident that something extraordinary is happening — in fact that man-made climate change is here. As well as those who claim they cannot yet see a definite link, other scientists, who accept that the enhanced greenhouse effect is already manifesting itself, say it is "up to us to show what is happening and up to politicians to decide what to do about it". Both lots of boffins, as the tabloid press used to call them, seem to me to be passing the buck. As a journalist who has been covering this issue for more than 20 years it has been my training to state the facts, record the debate about whether climate change is real, and let the readers make up their minds on the evidence presented. That remains the correct approach for the news pages of a newspaper. For a scientist that is also the right way to do things, at least when reporting new discoveries or theories in the scientific journals. But surely both for scientists and the rest of us responsibility does not stop there.

This is because as time has passed it has become increasingly clear that the matter is

Above: Bryn Mawr Glacier, Prince William Sound, Alaska, one of the many vast ice rivers which are now collapsing in the Arctic regions as temperatures rise faster than elsewhere. Glaciers are not just getting shorter, they are also getting thinner, releasing more and more water into the oceans and adding to sea level rise. Only in Scandinavia, where precipitation has increased and heavier snow is falling in the mountains, are the glaciers apparently growing. Elsewhere across the mountain ranges of the world, and near both poles, glaciers are shrinking and in some places disappearing completely.

Above: Pancake ice is soft newly formed sea ice that heralds the beginning of winter in the Ross Sea, the nearest open water to the South Pole. Within days of pancake ice appearing the whole sea surface freezes solid and any iceberg or ship caught in it will remain locked fast until the following spring, as early explorers discovered. The sea ice is an essential breeding ground and protection for tiny sea creatures like krill, the main food sources of many species of whale and penguins. The extent of sea ice loss each winter is leading to a reduction in the krill and a crash in populations of penguins because they are not able to rear their young.

Above: Icebergs are formed when glaciers or ice-shelves break off into the sea. Some are vast and take many years to break up and melt. They are a store of fresh water and change the salt content of sea water, which can in turn affect currents like the Gulf Stream. This is a natural process but it has speeded up both in the Arctic and Antarctic.

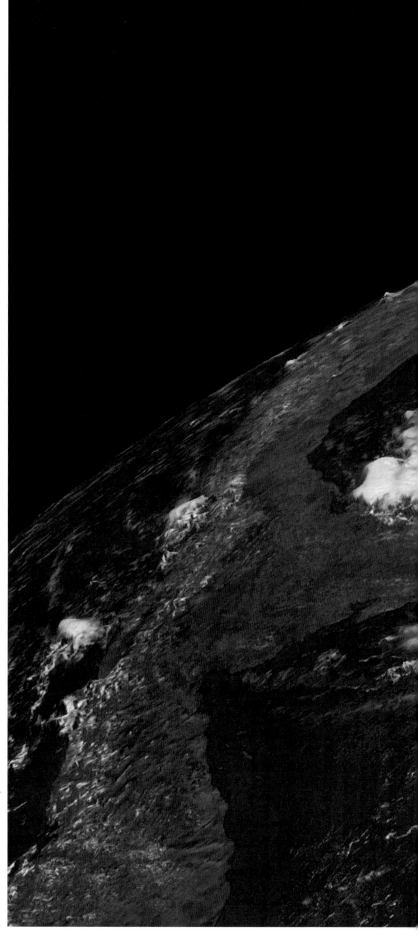

"After the tragedy of hurricane Katrina, many Americans now believe that we have entered a period of consequences... they are beginning to demand that the administration open its eyes and look at the truth, no matter how inconvenient it might be for all of us – not least for the special interests that want us to ignore global warming.

"The climate crisis may at times appear to be happening slowly, but in fact it is a true planetary emergency. The voluminous evidence suggests strongly that, unless we act boldly and quickly to deal with the causes of global warming, our world will likely experience a string of catastrophes, including deadlier hurricane Katrinas in both the Atlantic and Pacific."

Al Gore in his essay
'The Moment of Truth', Vanity Fair
April 17th, 2006.

Right: The awesome sight of hurricane Katrina seen from space, in this computer-enhanced image, as it approaches the Mississippi delta from the south-south-east on August 28th, 2005, just before it struck New Orleans. The storm gained in strength from a category 1 hurricane to a maximum 5 as it crossed the Gulf of Mexico, where in 2005 the sea water was the warmest ever recorded, giving Katrina more energy to draw on. The low pressure in the centre of a hurricane creates a dome of water in the ocean, which in this case overwhelmed coastal defences causing serious flooding. Although New Orleans was warned days in advance of the approach of the storm, and large numbers of the citizens had headed north to escape the devastation, many poorer people, without transport, remained behind and were caught by the fierce winds, driving rain and the floods, when the inadequate levees collapsed under the weight of water.

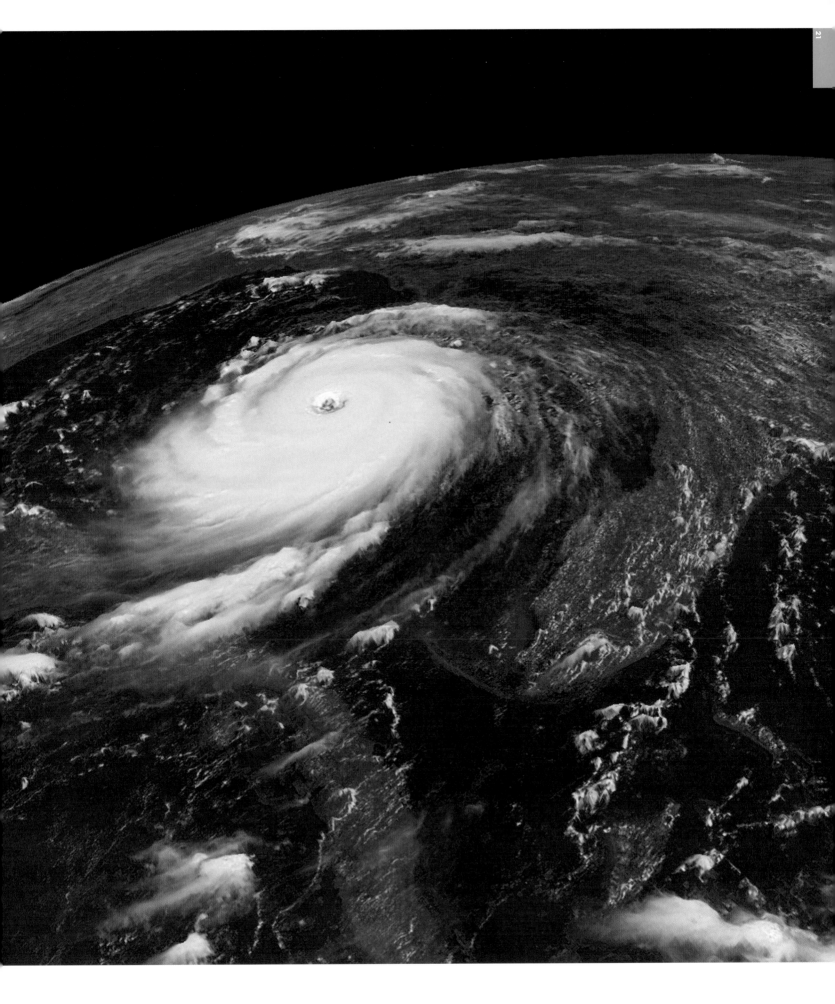

Above: The body of a victim of hurricane Katrina lies in the remaining floodwaters and debris in the Saint Bernard area of New Orleans two weeks after hurricane Katrina hit the city famous for its jazz. Images of bodies left uncollected amid the devastation of an affluent tourist city shocked the people of the richest country on earth, and the rest of the world. Because much of the city lies below sea level the water could not drain away and had to be pumped out of the suburbs, delaying emergency services and evacuation.

Above: Stung by criticism of the failure of America to cope with the damage caused by hurricane Katrina, and the administration's inability to look after the thousands of people left stranded without food and water, the US president, George W Bush, toured Louisiana by helicopter. As he flew over New Orleans five days after the storm he could see teams of rescuers still struggling to reach people in need of evacuation, and looters raiding abandoned businesses and homes. Bush said it would take years to recover from the devastation to Alabama, Mississippi and Louisiana.

Right: The force of the hurricane and the giant waves whipped up by the winds damaged oil platforms, refineries and installations all along the coast. These yachts were picked up by the flood water from a marina near New Orleans — just a small part of the severe property damage along hundreds of miles of coast.

"The monumental ruins left behind by… past societies hold a romantic fascination for all of us. We marvel at them when as children we first learn of them through pictures… We feel drawn to their often spectacular and haunting beauty, and also to the mysteries that they pose… Lurking behind this romantic mystery is the nagging thought: might such a fate eventually befall our own wealthy society? Will tourists someday stare mystified at the rusting hulks of New York's skyscrapers, much as we stare today at the jungle-overgrown ruins of Maya cities?"

Jared Diamond, 'Collapse: How Societies Choose to Fail or Survive', 2005.

Above: Three civilisations that once had the time and resources to build great monuments — Easter Island in the Pacific (top), the Mayan city of Palenque in Mexico (centre), and Ur in Sumeria in modern Iraq (bottom) — have faded away. History tells us that only when people have plenty to eat and have met all their basic needs do they have the leisure to construct symbols of their power and wealth. Why civilisations collapse is just as interesting, and the main cause seems to be overuse of scarce resources, exactly what is happening across the world in the 21st century. Archaeologists are still puzzling about why the Mayan civilisation collapsed between AD 731 and 774, but in the case of Easter Island and Sumeria the answers are known. The people of Easter Island cut down all their forests and so were unable to build the canoes in which they hunted for dolphins and other sea food which was their staple diet. Both the population and statue cult collapsed. This Sumerian ziggurat at Ur in Iraq being overflown by a US plane was abandoned 4,000 years ago when mismanagement of the irrigation led to soils becoming salty, food shortages, and the creation of a desert.

urgent. The scientists had hoped that someone, somewhere, would do something — but this hope has not been realised. It is surely now time to take a step further. Scientists who are still hiding behind the uncertainties, and the belief that they can do nothing more because it is someone else's job to act on the dangers they have revealed, should think again. So should the rest of us. As time has passed I have come to the view that waiting for someone else to act is no longer enough. Scientists and the rest of us have an ethical and moral dimension to our roles in society. We need to consider that time is short. Scientists are important because they are a group of people who have enough authority to wake the world up to the plight it faces. All of us need to play a part, and this book tells us how we can help.

Having got that out of my system it is fair to say that Europe, its peoples and political leaders are the most aware of climate change and its potential impacts. It is not clear why this should be so; perhaps it is because environment groups are most active, education levels are high, and the science base so strong. Europe is credited with being the world leader in action on climate change. Tony Blair, the UK's prime minister, was in 2005 at his most influential on the international stage. During his presidency of G8 he made climate change a priority for the summit. To set the scene he organised a conference to review the science. The news was shockingly bad. In response the political leaders made fine speeches, but in the event took no actions to make any significant dent in the problem. Carbon dioxide emissions continued to rise alarmingly across the world.

It is true that in February 2005, after years of uncertainty, the Kyoto Protocol finally came into force and 34 countries accepted legally binding

Above: Japanese children hold up a "planet of life" made of Shibori-zome, a silk tie-dye method traditional in Japan's ancient capital of Kyoto on December 2nd, 1997. It was made for display during the city's hosting of a crucial global warming conference, which led to the Kyoto Protocol, the first legally binding treaty to provide targets for industrialised countries to reduce greenhouse gas emissions by 2012. The "planet", a 2.5 metre diameter balloon covered in the expensive fabric, expresses the children's hopes for blue skies and seas and rich land for a clean and beautiful earth to be passed on to later generations.

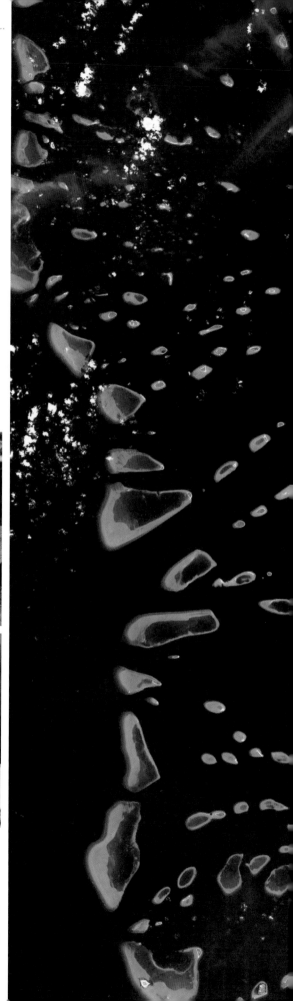

"As for my own country, the Maldives, a mean sea level rise of 2 metres would suffice to virtually submerge the entire country of 1,190 small islands, most of which barely rise over 2 metres above mean sea level. That would be the death of a nation."

Maumoon Abdul Gayoom, president of the Republic of Maldives, at the United Nations general assembly, New York.
October 19th, 1987.

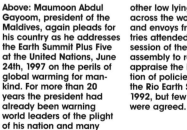

Above: Maumoon Abdul Gayoom, president of the Maldives, again pleads for his country as he addresses the Earth Summit Plus Five at the United Nations, June 24th, 1997 on the perils of global warming for mankind. For more than 20 years the president had already been warning world leaders of the plight of his nation and many other low lying island states across the world. Leaders and envoys from 173 countries attended this special session of the general assembly to review and appraise the implementation of policies adopted at the Rio Earth Summit in 1992, but few new actions were agreed.

Top: The Maldives is not just a tourist trap. Like any nation it has its own culture and traditional skills. In this case boat-building for the fishing industry. What will happen to this man and his craft skills when he has to move to another country and probably has to make a living miles from the sea?

Above: The Maldives looks a wonderful place for children to grow up. But these children face an uncertain future because as sea levels rise their homeland will disappear. By the time they reach middle age the entire nation will be faced with evacuation to an as yet unknown land, and a loss of their culture and heritage.

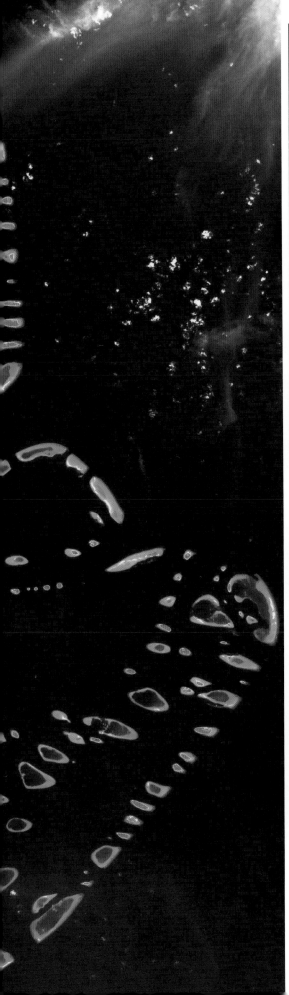

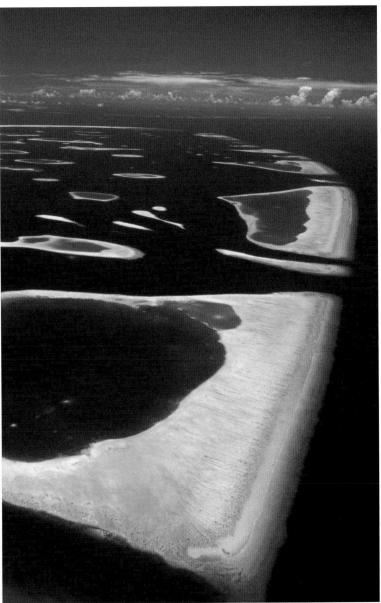

Left and above: North and South Malosmadulu Atolls, part of the Maldives, a republic in the northern Indian Ocean, southwest of India, with a population of 330,000 spread across a chain of 1,192 small coral islands with an average elevation of 1 metre above sea level. Waves triggered by the great tsunami of December 2004 spilled over the islands giving a foretaste of what will happen as sea levels continue to rise. Some populated islands were so damaged that they were evacuated and others became mere sand bars in the sea. From space and aircraft it is possible to see how vulnerable the islands are, spread out on coral reefs in the India Ocean. At ground level the Maldives are still the idyllic islands that so many thousands of tourists visit every year without giving a second thought to what an uncertain future faces their dream holiday destination.

"Earth provides enough to satisfy every man's need, but not every man's greed."
Mahatma Gandhi

targets for reducing carbon dioxide emissions. Later in the year at Montreal the world's nations met to finalise the details of the protocol and decide what to do to reduce emissions beyond 2012 when Kyoto expires. They decided to have another meeting in 2006 to talk about further talks. One of those meetings took place in New Zealand in April 2006 when Tony Blair was again on his feet talking about the dangers of climate change. In a revealing moment he told the conference that asking for action on global warming from politicians was the "purest example" of the clash between "long term interests and short-term gain." He said: "Often we are in a situation where it is not that governments do not want to do the right thing but that they worry electorally about the short-term consequences of doing so." Neatly put, but it also shows clearly that our current crop of politicians are not statesmen.

So there is a large and ever greater divide between what the scientists are saying, clearly spelling out dangers for the future, and the politicians' feeble policy responses. In the following pages some of the reasons for why this has happened in the past are explored. Of particular influence and interest is the way the fossil fuel lobby has succeeded in looking after its own interests at the expense of the planet.

But all is not yet lost. There are signs that at all levels of society — in business, in communities and through committed individuals — there are changes taking place. All over the world city mayors, local councils and their voters are taking the threat of climate change seriously. Under the Kyoto Protocol trading schemes designed to cut carbon emissions are under way. New clean technology is being transferred to people who would otherwise never have had the opportunity to develop without fossil fuel.

The Europe-wide carbon trading scheme, where companies can make money by selling tonnes of carbon they have saved through energy efficiency or installing cleaner plant, is making an impact on emissions. Across continents, there are ingenious new ideas, inventions and machines, which are being developed and need to be made universally available. In other places old, well proven ideas and methods, almost lost in the headlong rush towards a universal consumerist, throwaway society, are being revisited and revised to suit new conditions.

Some of these heartening developments are reviewed in these pages and their potential examined. As energy prices rise, and the oil begins to run out, as seems likely before the end of this decade, many of these technologies will be given a boost. New industries will boom as a variety of renewable technologies become heavily in demand to solve the looming energy crisis. This is good news both for society and the future of the planet. There is great potential for new jobs, and export industries for those with vision, as the wind turbine boom begun in Denmark has already shown. One of the great developments of this century will be small scale energy production in the home with wind, solar and hydro stations all grabbing a share of the market.

There is no single magic bullet to solve the coming energy crisis, and the danger to civilisation and the natural world that climate change represents. But society and individuals can encourage change and create a political will for action by buying into these technologies. It requires individuals at all levels to take responsibility for their action. At a personal level it matters what car you drive, what holidays you take, even whether you use energy-saving light bulbs (which we should all

Right: Dhaka, the capital of Bangladesh, was laying claim to be the most polluted city in the world before 2002. More than 50,000 two-stroke 3-wheeler taxis or rickshaws, known locally as baby taxis, were spewing black toxic smoke onto the streets. Levels of pollutants reached six times World Health Organisation maximum levels for lead, carbon monoxide and deadly particles that lodge in the lungs. A government ban on two-stroke vehicles in 2002, replacing them with four-stroke "green" taxis that run on compressed natural gas, has improved the situation, but traffic congestion and pollution are still chronic.

be doing already). But individuals are also savers, shareholders, businessmen, bankers, architects, scientists and politicians. Where do investors put their money, in clean or dirty industries? When consumers buy products, like wood or fish, they must demand to know that they come from sustainable sources. Scientists can no longer continue to hide in their ivory towers. If it is too much to hope that politicians can be persuaded to care even if the next election does not depend on it, can voters persuade them that green policies count at the ballot box too? Writing to your member of parliament and your local paper about your concerns is always a good start.

Perhaps what is most surprising, and cheering, about the battle to save the planet from man-made climate change is that there are people at every level of society committed to the cause. And this is not just in the old industrial world where wealth has already been created at the expense of the environment.

The accepted wisdom in America and most of Europe is that in countries like India and China the chief concern is economic development and environment comes a distant second place. But while this has been partly true, both countries now want clean development and are beginning to make it a priority.

For example in China, following massive dust storms, record vehicle pollution and grit from building sites which choked Beijing in April 2006, the prime minister, Wen Jiabao, ordered a re-think. The 30% increase in hospital admissions because of the smog, and the warnings that children needed to be kept indoors for their own safety, led the prime minister to say that the policy of economic growth at all costs had been wrong. He said

measures to conserve nature and improve air and water quality over the previous five years had not been successful, and had resulted in severe ecological degradation.

The government had failed to reach eight of its 20 environmental goals. From 2006, every six months, local government would be obliged to release information on energy consumption, which must be cut by 20% by 2011, and emissions of polluting chemicals, which must be reduced by 10% over the same period. The country needed to focus on the consequences of growth, he said.

In India the government is active in reducing air pollution in some cities. In New Delhi, the capital, where 70% of the pollution came from traffic, and levels of dangerous particles in the air were 10 times World Health Organisation limits, the supreme court ruled in 2003 that the entire public transport fleet must be run on clean fuels to protect the population. As a result 80,000 vehicles, including 9,000 buses and 45,000 auto-rickshaws, have been converted to run on natural gas. There was a dramatic reduction in carbon dioxide and other polluting emissions, although the growth of private vehicles threatens to wipe out the improvement.

But in India industry is also taking a stand. An example is the Tata Group, one of the continent's largest, oldest, and most respected business conglomerates, still run by the Tata family but with 2 million shareholders. It has 93 companies across the world and an annual turnover of $21 billion (more than £11 billion), but a cornerstone of its business is sustainability and holding back climate change. In the vibrant English that is a hallmark of India, here is a refreshing summary of Tata's philosophy.

Top: Switzerland is in the front line of climate change effects, with winter warming in the Alps damaging its vital skiing industry. The country has reacted by investing in alternative energy, as this photovoltaic electricity production plant on top of an office building in Bern shows.

Above: The G8 group of industrialised countries, which so far have talked a lot and done little about climate change, confronted by Greenpeace volunteers in Trieste, Italy in 2001. Fourteen activists, each from a different country, demand clean energy.

Right: Middlegrunden, off Copenhagen in Denmark, became the world's biggest offshore wind farm in 2001 with this impressive curve of turbines stretching 3.4 km (two miles) across the shallow sea not far from the city. The 20 two-megawatt turbines provide 3% of the Danish capital's electricity. Since this wind farm was built turbines have

increased in size and are soon each expected to have an output of five megawatts. The pioneering Danes are exporting the technology, supporting 20,000 jobs and creating one of the country's largest and most profitable industries.

"How much longer must we sit by and watch an increasing number of unusual natural disasters unfold before we realise that something is amiss? The small island countries warned more than a dozen years ago that there would be an increased frequency and intensity of storms and other adverse weather events as a consequence of global warming. In a sense they have served as mankind's early warning system… the canaries in the cages in the coal mines. Is anyone listening?"

Robert Van Lierop, writing in 2004 for the Institute of the Black World. He was one of the pioneers in 1990 negotiating the text of the Climate Change Convention.

"If there is anything at all we can assume about the shape our fragile world is taking, it is this: the condition of the air we breathe, the water we drink and the land we live on will all get worse if human rapaciousness continues to go unchecked.

"A big chunk of the responsibility for containing the plague driving our polluted and populous planet towards peril rests with industry and business. Balancing the imperatives of creating jobs and selling products and services with the absolute necessity of protecting and regenerating what remains of the natural environment is an onerous challenge. That it can be done is beyond doubt, but this is a task requiring a commitment to ideals more than bottom lines, to the good earth rather than profiteering."

In line with modern environmental thinking there is nothing in Tata's anxiety about the future that implies a reduced quality of life; in fact, the opposite. The company's plan is that both the existing generation and the next can enjoy this rich world, rather than watch powerless as climate change takes hold. As we can see in the final chapters of this book communities across continents are already making painless changes to wean themselves off fossil fuels for cooking, heating and lighting, and provide themselves with a clean water supply and sanitation. As they do so they improve their own quality of life. In the manufacturing sector the race is on to exploit clean technologies and create new industries and thousands of jobs. In older industries, for example car manufacturing, the competition is focused on finding ways to mass-produce vehicles that do not rely only on oil for propulsion. But much more needs to be done, and quickly.

With international political leadership still going far too slowly to solve the problem, it is up to committed individuals at city and community level to make the difference. Increasingly it is clear that personal choices need to be made about how we could each lead our lives to reduce our carbon footprint. It is not always easy while the world is still hell bent on an unsustainable path, but every day new products and new opportunities appear for individuals to help change the world.

The size of the problem is frightening but there is still time, just.

Left, top: Deserts can bury villages as the sand dunes move across the country like waves on the incoming tide. Here in Mauritania the women of Ljnanoune drag a giant net to the top of the encroaching dunes to try and hold back the Sahara desert from smothering their homes and fields.

Left, bottom: The rapid retreat of the Gurschen Glacier in the Andermatt region of Switzerland led the ski resort to cover ice with a specially made fleece at the start of the 2005 summer in a bid to cut the rate of melting by blocking out the ultraviolet rays of the sun. The shrinking of the glacier has meant that the resort, which

attracts 250,000 visitors annually, has had to build a larger ramp each year to get skiers to the slopes. The 4,000 sq metre fleece is designed to cut ice loss by 75%. The alpine glaciers are losing on average 1% of their mass each year threatening many of the ski resorts with bankruptcy.

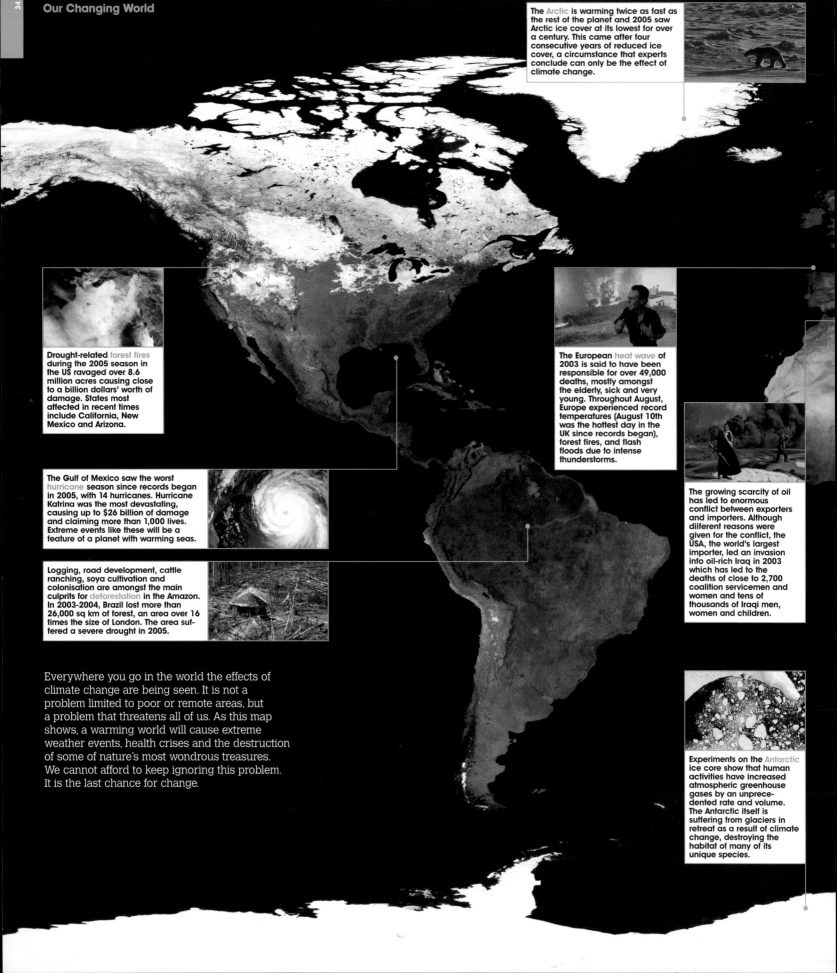

The Arctic is warming twice as fast as the rest of the planet and 2005 saw Arctic ice cover at its lowest for over a century. This came after four consecutive years of reduced ice cover, a circumstance that experts conclude can only be the effect of climate change.

Drought-related forest fires during the 2005 season in the US ravaged over 8.6 million acres causing close to a billion dollars' worth of damage. States most affected in recent times include California, New Mexico and Arizona.

The Gulf of Mexico saw the worst hurricane season since records began in 2005, with 14 hurricanes. Hurricane Katrina was the most devastating, causing up to $26 billion of damage and claiming more than 1,000 lives. Extreme events like these will be a feature of a planet with warming seas.

Logging, road development, cattle ranching, soya cultivation and colonisation are amongst the main culprits for deforestation in the Amazon. In 2003-2004, Brazil lost more than 26,000 sq km of forest, an area over 16 times the size of London. The area suffered a severe drought in 2005.

The European heat wave of 2003 is said to have been responsible for over 49,000 deaths, mostly amongst the elderly, sick and very young. Throughout August, Europe experienced record temperatures (August 10th was the hottest day in the UK since records began), forest fires, and flash floods due to intense thunderstorms.

The growing scarcity of oil has led to enormous conflict between exporters and importers. Although different reasons were given for the conflict, the USA, the world's largest importer, led an invasion into oil-rich Iraq in 2003 which has led to the deaths of close to 2,700 coalition servicemen and women and tens of thousands of Iraqi men, women and children.

Experiments on the Antarctic ice core show that human activities have increased atmospheric greenhouse gases by an unprecedented rate and volume. The Antarctic itself is suffering from glaciers in retreat as a result of climate change, destroying the habitat of many of its unique species.

Everywhere you go in the world the effects of climate change are being seen. It is not a problem limited to poor or remote areas, but a problem that threatens all of us. As this map shows, a warming world will cause extreme weather events, health crises and the destruction of some of nature's most wondrous treasures. We cannot afford to keep ignoring this problem. It is the last chance for change.

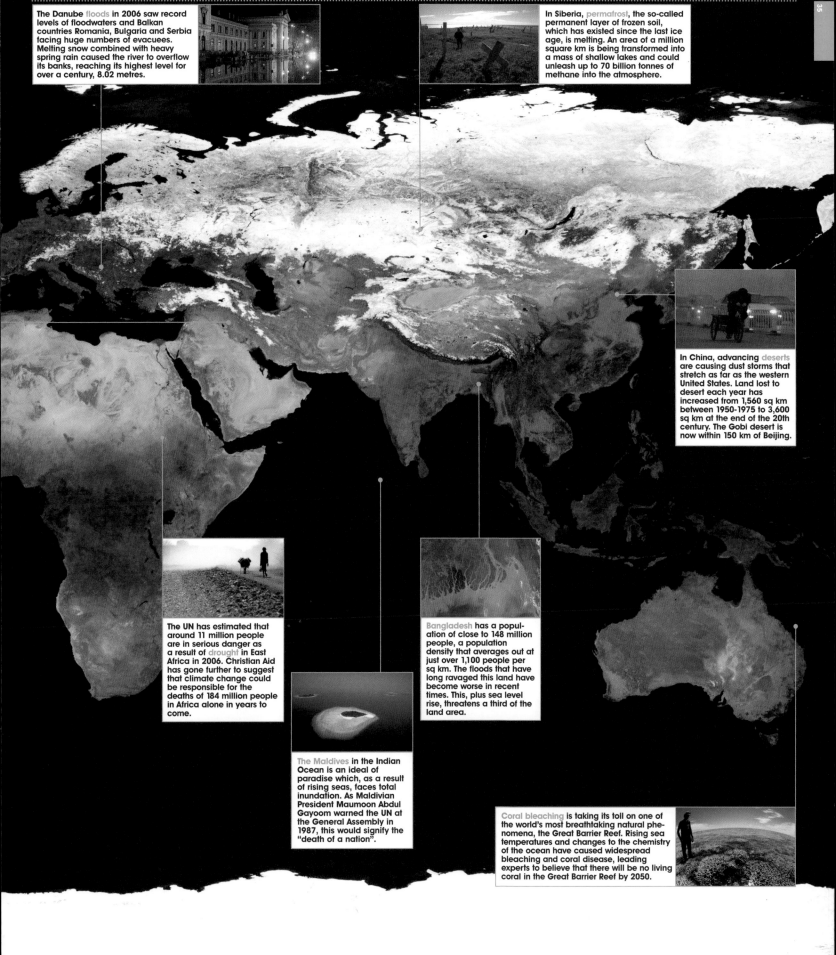

The Danube floods in 2006 saw record levels of floodwaters and Balkan countries Romania, Bulgaria and Serbia facing huge numbers of evacuees. Melting snow combined with heavy spring rain caused the river to overflow its banks, reaching its highest level for over a century, 8.02 metres.

In Siberia, permafrost, the so-called permanent layer of frozen soil, which has existed since the last ice age, is melting. An area of a million square km is being transformed into a mass of shallow lakes and could unleash up to 70 billion tonnes of methane into the atmosphere.

In China, advancing deserts are causing dust storms that stretch as far as the western United States. Land lost to desert each year has increased from 1,560 sq km between 1950-1975 to 3,600 sq km at the end of the 20th century. The Gobi desert is now within 150 km of Beijing.

The UN has estimated that around 11 million people are in serious danger as a result of drought in East Africa in 2006. Christian Aid has gone further to suggest that climate change could be responsible for the deaths of 184 million people in Africa alone in years to come.

Bangladesh has a population of close to 148 million people, a population density that averages out at just over 1,100 people per sq km. The floods that have long ravaged this land have become worse in recent times. This, plus sea level rise, threatens a third of the land area.

The Maldives in the Indian Ocean is an ideal of paradise which, as a result of rising seas, faces total inundation. As Maldivian President Maumoon Abdul Gayoom warned the UN at the General Assembly in 1987, this would signify the "death of a nation".

Coral bleaching is taking its toll on one of the world's most breathtaking natural phenomena, the Great Barrier Reef. Rising sea temperatures and changes to the chemistry of the ocean have caused widespread bleaching and coral disease, leading experts to believe that there will be no living coral in the Great Barrier Reef by 2050.

How Close is Runaway Global Warming?

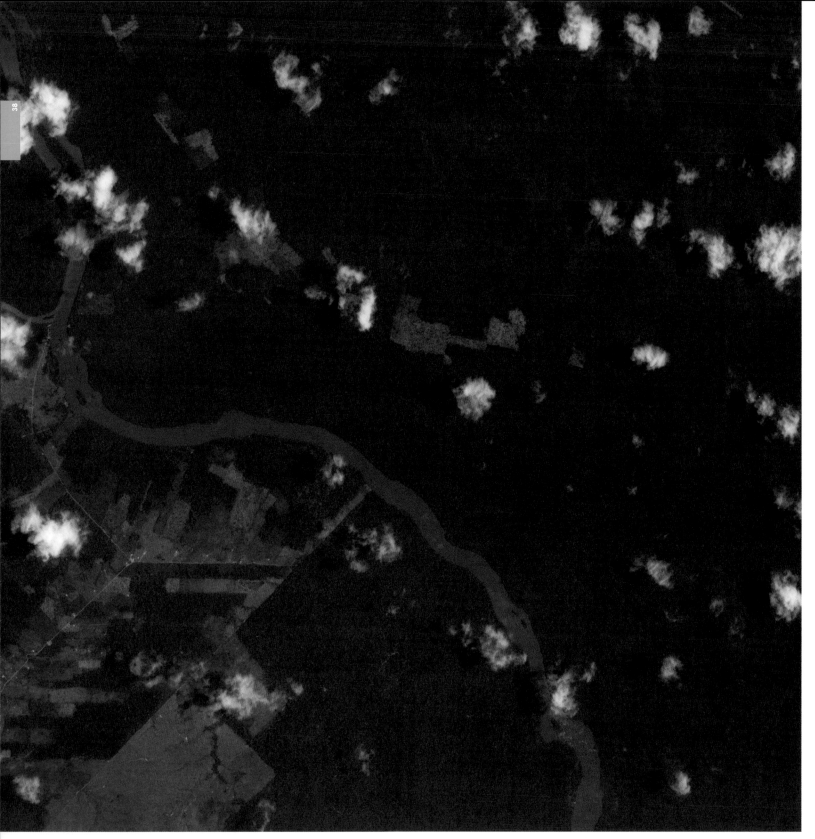

Previous spread: Teenagers run from waves breaking over sea defences in the northern Taiwanese city of Keelung on October 16th, 2001. It was a dangerous adventure. This typhoon, called Haiyan, was the ninth storm to hit the island of Taiwan in that year. The violence of the typhoons broke records for the amount of damage caused to Taiwan and and number of lives lost. All over the world the violence of storms is increasing, in line with the predictions of scientists using sophisticated computer models to look at future weather patterns.

Above: Roads driven through the Amazon basin inevitably bring human settlement and development. This high resolution satellite image of land either side of a tributary of the Amazon shows farms cleared from the jungle spreading on either side of several new tracks. Where there are no roads to the north of the river the jungle is virtually unbroken. The river itself is coloured brown by sediment, possibly from deforestation upstream. The picture is part of a scientific study of the region to see the effects of human-caused change in the region, and the possible effects on climate.

Above: The stark contrast between virgin jungle and land cleared from farming in the Mato Grosso state of Brazil, one of the greatest areas of forest destruction in the Amazon basin. The soil is poor and once cleared of trees rapidly loses its moisture and fertility and in some areas has already turned into semi-desert. Attempts by the Brazilian government to stem the tide of destruction have so far failed, since many of the frontier regions of this vast country remain lawless and corrupt.

The slow political reaction to the increasingly dire warnings of scientists about the fate of the earth has left many who have been following events shaking their heads both in sadness and in disbelief.

The European Union, more prepared than any other power bloc to face the facts, has accepted scientific advice that any more than a 2°C rise in temperature above pre-industrial levels risks catastrophe. Above that temperature natural systems will find it hard to adapt. Many specialised species will be unable to thrive in a changed climate and complex relationships between plants, insects and animals will be thrown into chaos.

But worse than that, an increasing number of scientists believe that above that temperature climate change might get out of control, at least beyond the ability of mankind to control it. There are two different ideas here, runaway climate change and a so-called tipping point. The first term sounds a bit like science fiction, but this idea of runaway climate change is not new. It is what has happened on Venus, our nearest planet, where the atmosphere is untenable to life as we know it on earth.

Runaway climate change is a theory of how things might go badly wrong for the planet if a relatively small warming of the earth upsets the normal checks and balances that keep the climate in equilibrium. As the atmosphere heats up, more greenhouse gases are released from the soil and seas. Plants and trees that take carbon dioxide out of the atmosphere die back, creating a vicious circle as the climate gets hotter and hotter.

The second phrase, tipping point, is heard a lot more from scientists. This is where a small amount of warming sets off unstoppable changes, for example the melting of the ice caps. Once the temperature rises a certain amount then all the ice caps will melt. The tipping point in many scientists' view is the 2°C rise that the EU has adopted as the maximum

limit that mankind can risk. Beyond that, as unwelcome changes in the earth's reaction to extra warmth continue, it is theoretically possible to trigger runaway climate change, making the earth's atmosphere so different that most of life would be threatened.

As with a lot of climate science, what used to be theory is now being seen in practice on the ground. New information makes clear that reaching the tipping point is a much more immediate threat than was previously thought. Some of the causes of the deep concern that we might lose control of our own destiny are discussed below. The danger grows with the increase in average temperature above what is called the pre-industrial level — the mid-18th century. Some scientists estimate that when the temperature reaches an extra 2°C above that equilibrium the earth's natural systems will be in serious trouble. It will affect many species' survival prospects, including our own.

So the key question is how close are we to a 2°C rise, and when will we get there? The first thing to admit is that nobody knows for sure, but many who understand the science say the answer to this twin question is, first, that we are already very close, and second, we might get there terrifyingly soon. In fact the 2°C threshold is much closer than almost anyone outside the specialist scientific community is prepared to acknowledge. By any standard, if you care about the future of the human race, it is too close for comfort. So to the vital question of when we might reach 2°C above pre-industrial levels; in other words how much time do we have to curb our excess emissions? Warming is directly related to the quantities of greenhouse gases there are in the air, the chief of which is carbon dioxide.

Above: A forest fire arriving in the town of Malhao in the southern Portuguese province of Algarve, in 2004, one of dozens of fires that ravaged the countryside during droughts and heatwaves over a three year period. This resident was unable to save his home and runs for his life. Hundreds of firefighters supported by planes and helicopters, some sent from France and northern Europe, battled against the forest fires across Portugal in a heat wave that reached a high of 40°C. Portugal had already lost 13% of its forests and woodlands in 2003, before these fires caused further destruction.

"One person flying in an aeroplane for one hour is responsible for the same greenhouse gas emissions as a typical Bangladeshi in a whole year."

Beatrice Schell, European Federation for Transport and Environment, November 2001.

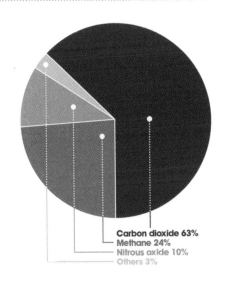

Carbon dioxide 63%
— Methane 24%
— Nitrous oxide 10%
— Others 3%

Although there are a number of greenhouse gases, carbon dioxide is the major contributor to warming expected over the next century. It is the burning of fossil fuels which is directly responsible for CO_2 increase in the atmosphere. It takes around 100 years for carbon dioxide to be absorbed back into living plants. Methane disappears more rapidly but still needs urgent controls because it is a more potent warming gas. It comes from agriculture, rotting rubbish dumps, oil and coal exploration and leaky pipelines.

Source: Hadley Centre for Climate Prediction and Research.

Concentrations of carbon dioxide in the atmosphere are already at 382 parts per million. That is up from the pre-industrial level of 280 ppm, a considerable increase. To get that in perspective we need to realise that the 280 ppm figure had remained more or less unchanged for 10,000 years, the period which accounts for the entire span of modern human history. The benign climate that has allowed the human race to multiply, develop and prosper has remained stable through that period. There have been minor variations: warm periods that allowed places like Greenland to be settled by the Vikings or mediaeval monks to make wine in Britain, and cold periods, known as mini-ice ages, that made it possible to have frost fairs on the frozen Thames in London during the 17th and 18th centuries. The last one was held in the winter of 1814. These so-called natural variations in the climate have allowed those trying to rubbish global warming theories plenty of ammunition. But those changes have now been well studied and are better understood. It is no longer credible to suggest that what is happening now is a natural variation of a sort recorded in the last 2,000 years. In fact the variations in the quantities of carbon dioxide in the atmosphere have been small in that period, and other natural variations like sunspots have been the culprits for the previous warm and cool periods. The recent increases in greenhouse gases have changed all the rules and the stability in the climate system man has enjoyed so long.

Current calculations suggest that if and when the level reaches 450 parts per million there will be a 50% chance of the earth's temperature exceeding a rise of 2°C — in other words an even chance of potentially catastrophic climate change. To be on the safe side (the so-called precautionary principle, which so many politicians claim they endorse) some scientists believe that the carbon dioxide in the atmosphere must be pegged back to 400 parts per million — a mere 18 ppm above the current level. So, on their current calculations, since man began the industrial revolution, and unwittingly an experiment with the climate, the human race has already got more than 80% of the way to causing a potential disaster.

On this evidence it is clear that drastic action is needed. Some scientists have certainly been urging politicians to take urgent and immediate action. Recent evidence demands, according to a consensus of the world's best climate scientists, that we need to cut existing emissions by between 60% and 80% in the next 40 years to stand a chance of preventing climate change becoming unstoppable, and keeping control of our own destiny. Compare that figure with that achieved by the Kyoto Protocol, to date the best effort by politicians to cut emissions. This will cut greenhouse gases from 34 of the developed countries by 5.2%, excluding the world's biggest polluter, the United States. Over the period of the agreement, which lasts only until 2012, total world emissions will rise because of the growing industries of the developing world.

What does the science tell us about how much time we have left to solve the problem? Measurements taken by NASA Goddard Institute for Space Studies and Columbia University Earth Institute New York, released in December 2005, show that in the last 100 years the world's average temperature has increased by 0.8°C. That seems to leave a comfortable 1.2°C to go before the tipping point is reached, but this is where the climate plays a nasty trick. Unlike glass in a greenhouse, the extra heat-trapping gases released into the air take time to build up their full effect. This is largely because

Left: A group of people sit on a sea wall watching the sunset over the Arabian Sea in Mumbai (Bombay) in March 2001. Pollution over the western Indian city, like many in the Indian sub-continent, has reached such extremely high levels that it is the cause of spectacular sunsets. Hundreds of people stop and sit to watch as they walk home from work along the city's harbour foreshore. Even though the sky is clear of clouds the sun can disappear completely before it reaches the horizon because the pollution is so thick.

of the delaying effect of the cool oceans as they catch up with the atmosphere. Best estimates are that there is a 25 to 30-year time lag between greenhouse gases being released into the atmosphere and their full heat-trapping potential taking effect. That wipes out any feeling of comfort. It means that most of the increase of 0.8°C seen so far is not caused by current levels of carbon dioxide but by those already in the atmosphere up to the end of the 1970s. Still worse, the last three decades have seen the levels of greenhouse gases increase dramatically. In this 30-year period the earth has seen the largest increase in industrial activity and traffic in history. This great burning of fossil fuels has also coincided with the mass destruction of rainforests. So on top of the extra heat we are already experiencing there is another 30 years of ever accelerating warming built into the climate system.

Using their best understanding of how the climate works and the largest computers in the world to create a picture of what is happening scientists believe that in those three decades enough extra gas has been released to raise the earth's temperature by as much as 0.7°C on top of the increases already measured. Add the 0.8°C already measured to the 0.7°C, because of the last 30 years of growing emissions, and the human race is already committed to a total rise of 1.5°C.

The full import of this needs a little more explanation. As has become increasingly clear over the last 15 years, the more you know the worse the prospects appear to be. In this case it is the annual increases of carbon dioxide in the air. When annual measurements began 50 years ago the average rise a year was around 1 part per million, sometimes dipping below, often slightly more. By the end of the 1990s the average increase had risen to 1.5 ppm. In the

first five years of this century it has exceeded 2 ppm twice (2.08 ppm in 2002 and 2.54 ppm in 2003) before dropping back to 1.5 ppm. In 2005 it rose more than 2 ppm again. Over 50 years the amount of carbon dioxide in the air, trapping heat, has been on a general upward curve and now appears to be accelerating.

So in terms of the potential for runaway climate change what does all that mean? A paper presented to the International Symposium on Stabilisation of Greenhouse Gases in the Atmosphere in Exeter, England in February, 2005 gave a stark warning. Malte Meinshausen of the Swiss Federal Institute of Technology said that to be "on the safe side", that is not to gamble with the future of the planet, it would be sensible not to exceed 400 ppm of carbon dioxide in the atmosphere. At the beginning of 2006 the level had already reached 382 ppm. At a minimum this figure is rising by an average exceeding 1.5 ppm a year so unless there is a drastic change in human behaviour or some other unknown factor the 400 ppm safety threshold will be exceeded within 15 years or perhaps even a decade.

And this only deals with carbon dioxide. This is the main greenhouse gas, and the most important, but others also make a significant contribution to warming the atmosphere, particularly methane and nitrous oxide. Levels of both these are also rising above pre-industrial levels. According to the World Meteorological Organisation, in 2004 methane was at 1,783 parts per billion (ppb) in 2004, and nitrous oxide at 318.6 parts per billion, the highest ever recorded. This is an increase on pre-industrial levels of 155% and 18% respectively, and an increase in a decade of 37 ppb and 8 ppb in absolute amounts.

Left: A dust storm in Beijing turns the sky an amber colour and reduces visibility to 500 metres. This picture was taken in March 2002. Dust storms are a regular feature of life in the capital of China as the deserts creep ever closer to the capital and are now around 150 km (100 miles) away. Extensive deforestation and desertification in northern China have fuelled the dust storms and the Chinese estimate that nearly one million tonnes of Gobi desert sand blows into Beijing each year. The authorities are hoping this will not happen during the 2008 Olympic Games in the city.

"We are upsetting the atmosphere upon which all life depends. In the late '80s when I began to take climate change seriously, we referred to global warming as a 'slow motion catastrophe' one we expected to kick in perhaps generations later. Instead, the signs of change have accelerated alarmingly."

Dr David Suzuki, chair of the David Suzuki Foundation, October 2005.

Scientists calculate that by adding the warming effect of these to carbon dioxide the equivalent warming effect is that of another 45 parts per million of carbon dioxide in the atmosphere. If this is correct, the world is already well over the 400 ppm level, so it is it almost impossible to avoid dangerous consequences.

Some scientists think this is too pessimistic a picture because the human race can also adapt to the changes that climate change will bring. Professor Sir David King, the UK government's chief scientific adviser, giving a talk to senior industrial figures at Reuters in London in 2006, said the current target was to keep below 550 ppm. Afterwards I challenged him about the figure; surely that was taking a gamble, and risking dangerous climate change? He conceded that 400 ppm was the "scientific ideal" but that politically it was not attainable. It was no use asking politicians to go for a figure they could not reach.

But if the figures given at the Exeter conference are right then Sir David was surely advising politicians to take a serious gamble with the future of the human race. In a subsequent interview with the BBC he clarified his position. Current British government policy was to cut carbon dioxide emissions by 60% by 2050 but this was based on keeping below 550 ppm in the atmosphere. He conceded that on the latest science that would take the temperature up 3°C by the end of this century, well above the "danger" threshold. His solution was to go for adaptation, building sea walls, moving cities, and changing food crops. "After all, we have nearly 100 years," he said. He conceded, however, that the developing world would be hard hit by this approach and that some countries, particularly small island states like the Maldives, would disappear altogether. The

latest report to the Hadley Centre for Climate Prediction and Research in Exeter said a 3°C temperature rise would cause a drop worldwide of between 20 million and 400 million tonnes in cereal crops and put about 400 million people at risk of hunger.

Despite these dire forecasts some optimists argue that even if the 400 ppm threshold is exceeded it would still be possible to reduce current emissions over the next 50 years or so to bring the levels back down towards 400 ppm or even below. That would give the atmosphere a chance to stabilise and avoid the ice caps melting. The kind of effort required to achieve this would be enormous. In fact huge strides in technology and political effort would be required to get the 60% to 80% reduction in man-made emissions over the next 50 years. This is, as has previously been said, the level of cuts scientists believe is the minimum required if we are to avoid "dangerous climate change".

There is more about the issue of dangerous climate change in the next chapter, because it is an important phrase in the political sense. All the governments that have signed up to the 1992 Climate Change Convention, and there are 180 of them, have signed up to taking measures to avoid "dangerous climate change". So far they appear to have avoided asking themselves what this actually means in practice. But on any calculation, whether politicians accept 400 ppm as a target, or even the much laxer and riskier 550 ppm, time for action is short. At best we have probably only until 2020 to make decisions and change the way we power the planet to give the human race a reasonable chance of avoiding dangerous climate change. On current progress that is a tall order. At first sight it is hard to understand why this potentially imminent economic and social

Right, top: A Kenyan man walks with his donkey carrying water after trekking 6 km (4 miles) to the only well with water in El Wak, 1,530 km (950 miles) from Kenya's capital of Nairobi, in December 2005. The Mandera district was littered with carcasses and criss-crossed by nomads hunting for water as the remote plains of northern Kenya suffered a

drought that killed hundreds of thousands of livestock. The area, which borders Somalia and Ethiopia, was already one of Kenya's poorest and most arid areas before the drought struck.

Right, bottom: In another continent the desert is also still advancing despite efforts to stop it. A Chinese primary school student surveys barren fields after taking part in an afforestation project at Huai Lai county in Hebei province, 70 km (45 miles) northwest of Beijing. The environment in Huai Lai county has been deteriorating so much that

parts of fields and several small mountains have been buried in sand as a result of desertification. Official statistics show that over the past decades, desertification has advanced and more than 40% of China's territory had been turned into desert, contributing to the dust storms that can reach as far as the United States. Before the 1980s,

China's desertified land expanded at an average rate of 1,560 sq km a year (625 sq miles). The figure rose to 2,100 sq km a year (840 sq miles) in the 1980s and to 2,460 sq km a year (980 sq miles) in 1994.

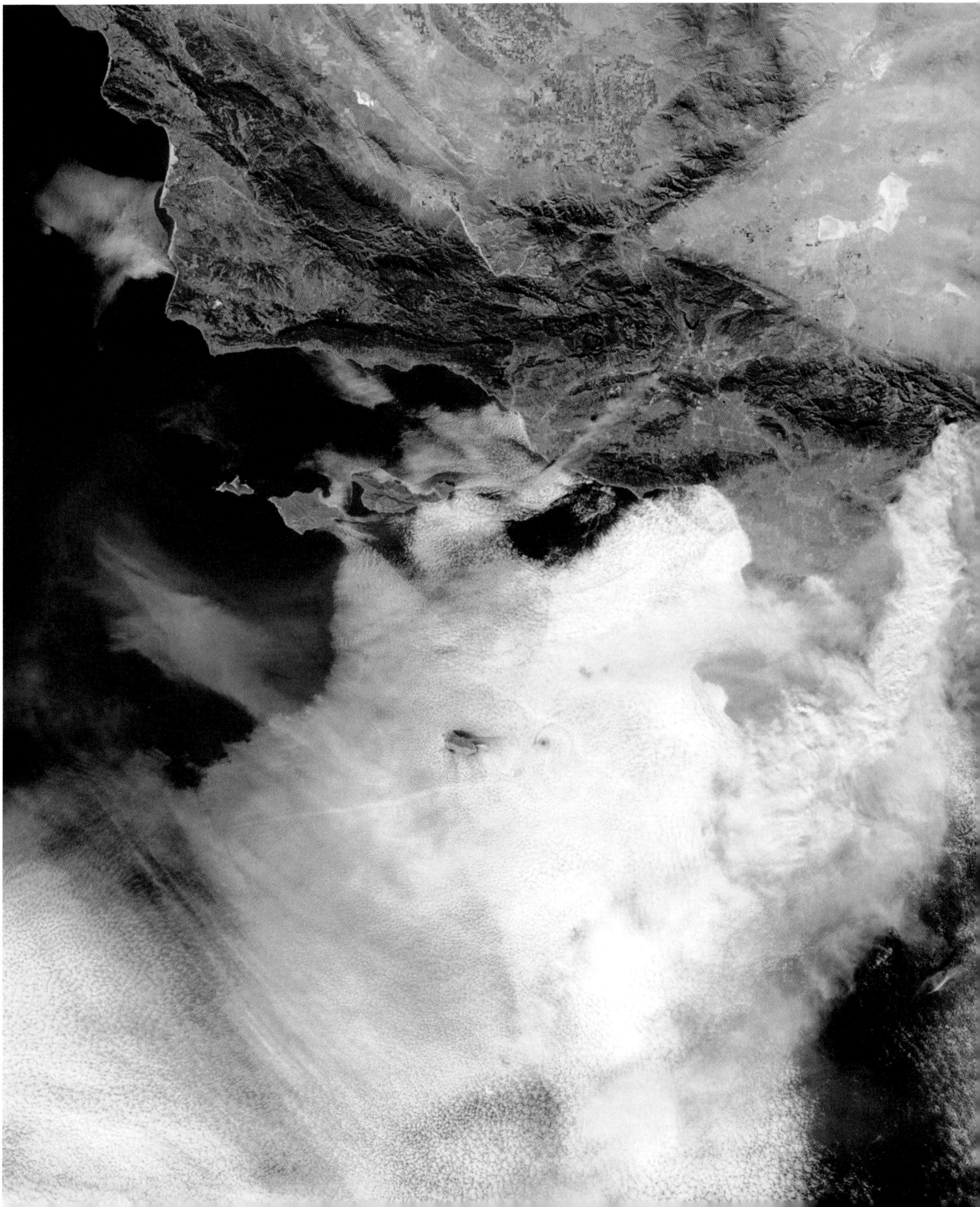

catastrophe is not discussed more. The science is not difficult to comprehend, the average 10-year-old could grasp the basics of it. Yet so shocking are the consequences that the standard response from those in power is to avoid the issue of how close we are to this dangerous threshold. Cynical observers might conclude that politicians while in office seem to fear that if they dwell on such a question then the consequences of the answer might limit their own survival. At first sight, at any rate, they will have to do something unpopular, something the public won't like. So they fear that if they protect the future of mankind the current generation will resent it and will vote them out of office. Although heads of state like to be called leaders, they appear to believe that voters, like the much maligned lemming, would rather run over a cliff than be told they have to stop running in order to avoid mass suicide. Experience of what is happening around the world, and there are many examples, show this is not so. Given the opportunity, the correct information, and access to the already proven technologies, both individuals and communities are prepared to work hard to combat climate change. It is also clear that if everyone was given the chance to be involved the problem could be solved.

The collective failure of the heads of the world's most important countries to lead their populations away from the abyss is examined in later chapters, along with ways in which the situation could be remedied. But to comprehend the extent of the leaders' dereliction of duty it is important at this stage for the reader to understand the science, which all the leaders been made aware of but would rather pretend they had not been. First of all there are lots of events that can make the climate warm faster and in a way that puts

Left: Forest fires in the hills above the suburbs of Los Angeles on October 25th, 2003 seen from a satellite. Huge wildfires are burning to the east of the city. Just one day later the situation was even worse with several massive fires raging across the region, driven by the fierce Santa Ana winds that blow towards the coast from the interior deserts.

Above: Los Angeles was one of the first cities in the world to realise that traffic was causing a major pollution problem when it became famous for its smogs in the 1970s, which were caused by the action of sunlight on exhaust fumes. This aerial fisheye lens view of Los Angeles at twilight shows part of the vast area the city covers and its network of highways. The problem of smog continues despite of the introduction of catalytic converters but is now compounded by the threat of global warming. In response the state has become a pioneer in encouraging low-emission vehicles and progressively tightening pollution controls despite fierce opposition from American car makers.

"America is still in denial about the energy problem and few politicians are prepared to accept painful solutions. We have a peculiar situation in which companies like General Electric and BP are more progressive on energy policy than the administration and Congress."

James Cooper, a Democratic congressman from Tennessee, July 26th, 2005.

Above: A demonstration of how reliant the modern world is on electricity. This scene is of a mass exodus of people and traffic leaving New York's Manhattan Island over the Brooklyn Bridge, when the city suffered a massive power cut on August 14th, 2003. Sweltering New York was thrown into chaos as thousands of commuters were stranded and many more trapped in the subway system. It brought back fearful memories of the September 11th attacks.

"When you start messing around with these natural systems, you can end up in situations where it is unstoppable. There are no brakes you can apply. This is a big deal because you cannot put permafrost back once its gone."

David Viner, a senior scientist at the Climate Research Unit at the University of East Anglia on the melting of the Siberian tundra, 2006.

reversing the process outside our control. There are also events that can act to cool the climate, again on a scale that dwarf man's attempts to prevent them. In the jargon they are called positive and negative feedbacks. An example of a positive feedback is melting ice and reduced areas of snow cover. Ice and snow, because they are white, reflect most light back into space. Bare rock, particularly dark rock, and seawater absorb much more of the sun's rays. You only have to sit on a stone wall after sunset and feel the heat on your behind to understand the process. Perhaps a better example is the contrast between black and white cars. A white car with the sun on it remains relatively cool compared with the heat absorbed by a black car. If the wall had been of white ice little or none of the heat would have been retained.

Over the last 30 years vast areas of ice in the Arctic and Antarctic have been disappearing. Rock in Greenland and the Antarctic that has been covered by permanent ice for 10,000 years is being exposed to sunlight in summer and is warming up. The same is true for even larger areas of the ocean around the North Pole. A second and related problem is the melting of the permafrost. Vast quantities of carbon are stored in the frozen soil, along with large amounts of methane from once rotting vegetation. As it warms, the stored carbon and methane is released to the air, increasing the warming effect. The destruction of forests also leads to ever larger amounts of stored carbon being released into the atmosphere as carbon dioxide. While most of this is man-made destruction, forests are also being damaged by climate change and releasing carbon dioxide as a result. The destruction of forests is estimated to add as much as one fifth of the carbon dioxide currently being added to the atmosphere by burning fossil fuels. Less tree cover also adds to

Left and above: The effects of thawing of frozen soil are hard to show on a photograph but have a profound effect on the locality and the climate. In the picture on the left taken in June 2002 the Siberian tundra was thawing out. It can be seen in the standing water lying in pools across the landscape. The tundra's permafrost, which is a frozen layer of soil many metres deep, which has been unchanged since the ice age, is turning into bogs, ponds, and wetlands. In this image, vegetation is bright green, and pooled water is dark blue or almost black, and can be seen in the upper left quadrant of the image in the Kolymskaya wetlands. Deep red areas show bare, exposed soil and in the upper right, this is where spring has yet to arrive in high-altitude terrain. All round the Arctic it is the same. On the ground in Alaska the effects of melting permafrost can be seen in what are called drunken forests, and graveyard crosses at odd angles. The ground heaves as it melts.

the problems of heat and drought, since trees retain moisture and temper the climate. All of these three positive feedbacks are already adding to the problems of climate change and are likely to accelerate the process. Examples of negative feedbacks, which are also difficult to measure, include what is known as the carbon dioxide fertiliser effect, industrial and vehicle pollution and cloud cover.

Plants and trees need carbon dioxide to grow through photosynthesis, so extra amounts of the gas in the air in theory have a fertiliser effect, making most things grow faster and so fixing more carbon. In laboratory conditions this can be measured precisely, for example by growing plants and trees in large greenhouses with increased carbon dioxide in the air. It works, they do grow faster. However, in the real atmosphere where all the other variables like moisture and temperature are not controlled, it is a far more difficult effect to measure.

Pollution — sulphur dioxide gas from coal fired power stations for example — reacts in the atmosphere to form tiny particles called aerosols, which reflect sunlight back into the atmosphere before it reaches the ground. The effect of aerosols, albeit from a different cause, has been measured in the spectacular aftermath of June 12th, 1991 when Mount Pinatubo in the Philippines erupted. It injected 20 million tonnes of sulphur dioxide into the atmosphere, along with huge quantities of dust. The clouds of material reached 19 km (12 miles) above the earth. Rather than falling back to the ground the plumes of the eruption were pushed so high that the material was caught up by the powerful horizontal winds that circulate in the stratosphere, which begins about 11 km (7 miles) above the earth. Instead of just having a local effect like smoke from a power station,

factory or city, the dust, particles smaller than one hundredth of a millimetre, spread round the world in a giant ring, causing among other things spectacular sunsets. The amount of radiation reaching the lower atmosphere fell by 2% as a result and the global average temperatures by 0.25°C for two years. By the end of that period the particles had fallen back to earth but for climate scientists it was a gigantic real life experiment. Another remarkable example occurred after the September 11th attacks in New York in 2001. For three days all air flights in the US were grounded and daytime temperatures increased, while night temperatures remained the same. Researchers concluded that lack of aircraft exhausts, an otherwise permanent feature of American skies, had made the difference. The tiny ice crystals in the exhausts normally reflect back the sunlight, an effect previously unrecorded. It has been called global dimming, but the effect of aircraft exhausts and other forms of pollution on clouds is extremely complex. Water vapour is a potent global warming gas but in the form of clouds it can act both to cool and to warm the climate.

Aerosols from pollution promote extra cloud cover. As anyone who has travelled in an aircraft knows, the dazzling white of the top of the clouds reflects sunlight back into space. Many aerosols have the effect of making clouds appear brighter from space. This has an additional cooling effect by reflecting back the sunlight. Research published in December 2005, jointly by the UK's Meteorological Office and the US government's National Oceanic and Atmospheric Administration, looked at the problem again and concluded that the cooling effect of aerosols from factory chimneys, forest fires and dust particles swept up by desert storms, which had previously been taken into

Right: Aircraft are the fastest growing contributor to global warming as air travel continues to increase. As well as greenhouse gases the water vapour causes these contrails, seen here over Deal in Kent, England, which help to form high cloud. Scientists are still working on exactly what effect this has on global warming. But the effect of aircraft pollution has been estimated to be about 2.7 times more than that of their carbon dioxide emissions alone because of where they are emitted into the atmosphere. Despite many calls for aircraft fuel to be taxed the United States has so far vetoed all proposals to do so.

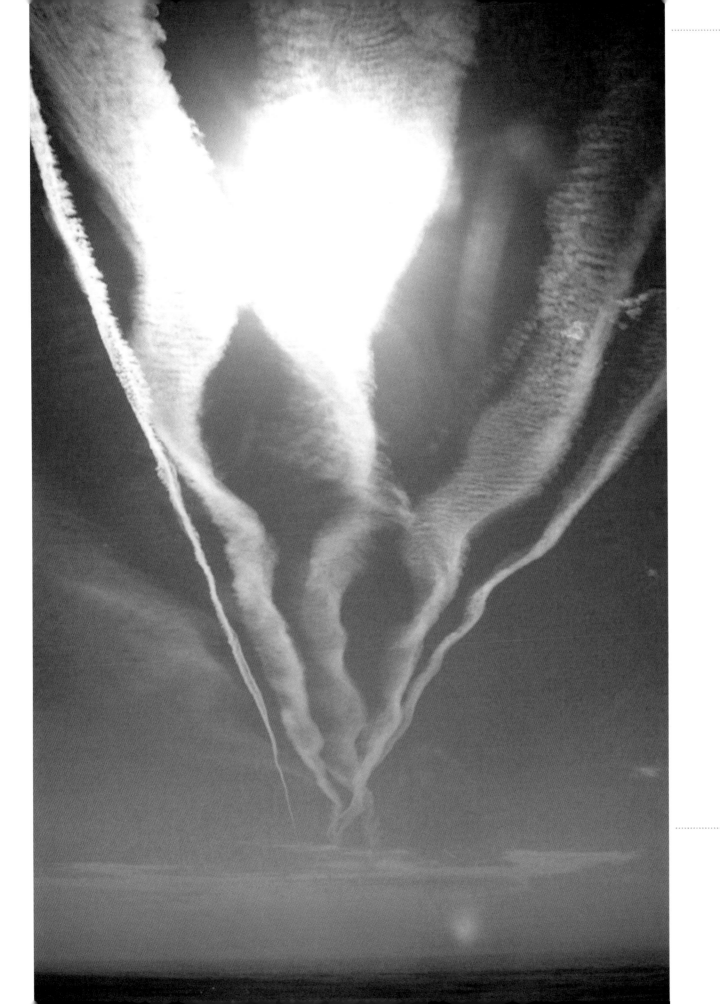

"The weight of evidence for climate change is very strong indeed, and has got stronger over the years since I began chairing the meetings of climate scientists in 1988. The rate of warming is now greater than it has been for 10,000 years, that means the rate of climate change is greater than it has been for 10,000 years."

Sir John Houghton, the first chairman of the UN's Intergovernmental Panel on Climate Change, 2006.

account and calculated, had been badly underestimated. Nicolas Bellouin, a climate modeller at the Met Office, summed up. "We found that aerosols actually have twice the cooling effect we thought. The consequence is that as air quality improves and aerosol levels drop, future warming may be greater than we currently think." Cloud also has another effect, acting like a blanket at night to keep the heat trapped at the earth's surface. It is no accident that in temperate regions clear nights in winter are frosty and cloudy ones keep the ground temperature well above freezing. Depending on the height of the clouds and the time of day or night the reflective or blanketing effect of clouds obviously varies hugely. Scientists are still studying this vital area and are trying to discover how different types of cloud at different heights affect temperature, and whether changes in cloud cover are keeping the earth warmer or making it colder. The latest information is that aviation is a far greater cause of climate change than was previously believed. Not only is it the world's fastest growing source of carbon dioxide emissions, aircraft also produce water vapour. This condenses to form ice crystals in the upper part of the lower atmosphere, known as the troposphere. These ice crystals, popularly known as vapour trails, trap the earth's heat. Taken together, the carbon dioxide, local ozone formation from other aircraft emissions, and the condensation trails of the aircraft have 2.7 times the effect of carbon dioxide alone.

These frontier areas of science and the uncertainties that arise because of changing cloud cover mean there is still plenty of room for doubt about how quickly temperatures will rise and how far. Nearly all the new evidence points to the fact that the rises will be greater than previously thought. One other vital point is

about changes in the measurements of global average temperatures. To the layman they can be misleading, because they appear quite small. What has become apparent is that the planet does not warm evenly. The places which are predicted to get warmer more quickly are the Arctic and the Antarctic, where vast quantities of ice, and in the northern hemisphere permafrost, dominate the landscape. In fact permafrost covers 27% of the earth's land surface. These are exactly the areas where the most dangerous positive feedbacks occur and the melting is already occurring. This is what worries scientists about the average of 2°C across the whole globe. Localised warming in Greenland will not be the average 2°C but is expected to be above 2.7°C, probably reaching 3°C. This rise is significant because it is above the point that triggers the melting of this vast ice cube. In melting, the 1 million square miles of two-mile-high Greenland ice cap will raise sea levels by seven metres (22ft). This is not all at once of course, but unstoppable once the tipping point is reached. Even though scientists believe it could take between 1,000 and 3,000 years before all the ice disappears the sea level rise will begin to have impacts much sooner.

In February 2006 the latest measurements of Greenland showed that its glaciers were unloading three times as much ice into the sea as in 1996. This meant that previous estimates of Greenland's contribution to sea level rise had to be revised. The rise due to Greenland's ice losses was thought to be tiny but current estimates have risen to half a millimetre each year. This may not sound much but to scientists it was very ominous. Liz Morris, Arctic science adviser at the Scott Polar Research Institute at Cambridge, said: "It appears Greenland is just on the turning point."

Right: Baffin Bay, located between Greenland and Baffin Island, is the site of noticeable climate change. Ever larger quantities of low salinity melt waters from the disappearing ice moving south into the warmer seas are affecting ocean currents. The flow rate of the Jakobshavn glacier on the western slope of Greenland has increased significantly and is now moving at about 10 km (six miles) per year, calving a large number of icebergs into Baffin Bay and adding to sea level rise.

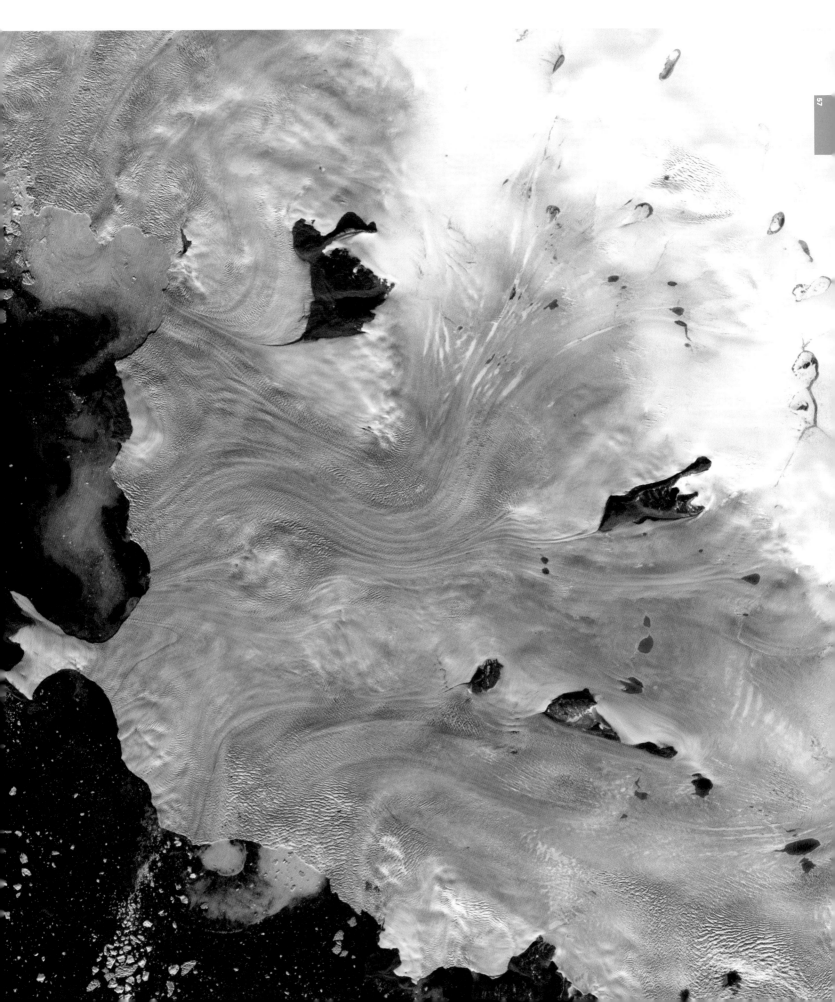

Creating a Scorched Earth

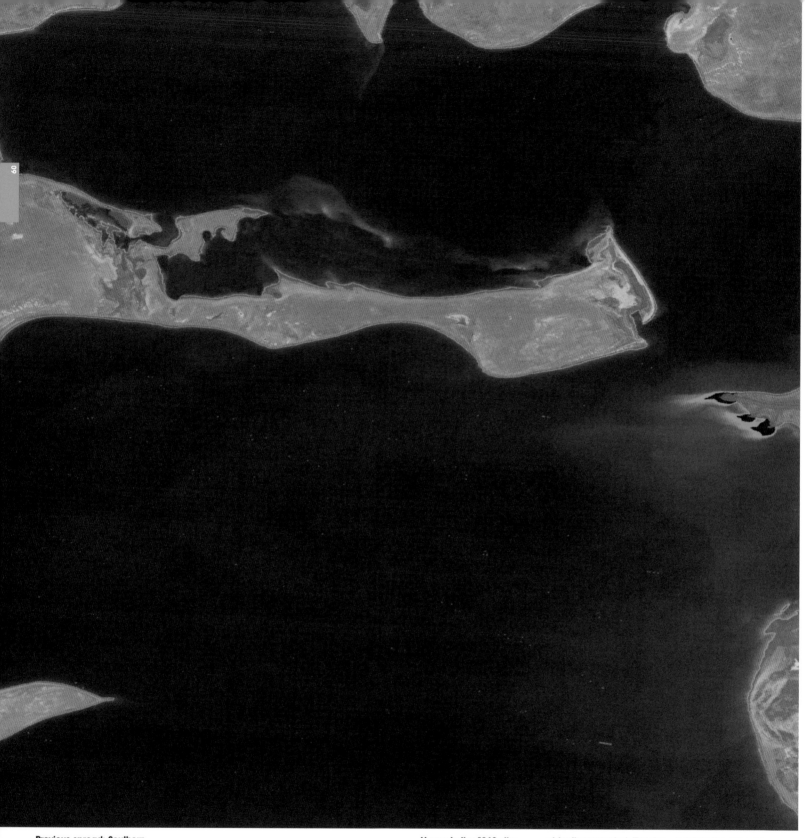

Previous spread: Southern Europe from Greece across to Spain and Portugal has been increasingly plagued by forest fires in the last decade as summers become hotter and drier. Here in August 2005 fire-fighters size up the task ahead as a wildfire burns a forest near the village of Bouleternère, near Perpignan, southern France.

Above: In the 1960s the Soviet Union began to divert the two rivers that fed the Aral Sea to provide irrigation so that cheap cotton could be grown in what was barren landscape. The sea, once the fourth largest lake on earth, began to shrink. This picture was taken in 1973. By 1989 the northern and southern parts of the sea had already become virtually separated. The drying out of the sea's southern part exposed the salty seabed. Dust storms increased, spreading the salty soil right into the agricultural lands. As the agricultural land became contaminated by the salt, the farmers tried to combat it by flushing the soil with huge volumes of water. What water made its way back to the sea was increasingly saline and polluted by pesticides and fertiliser.

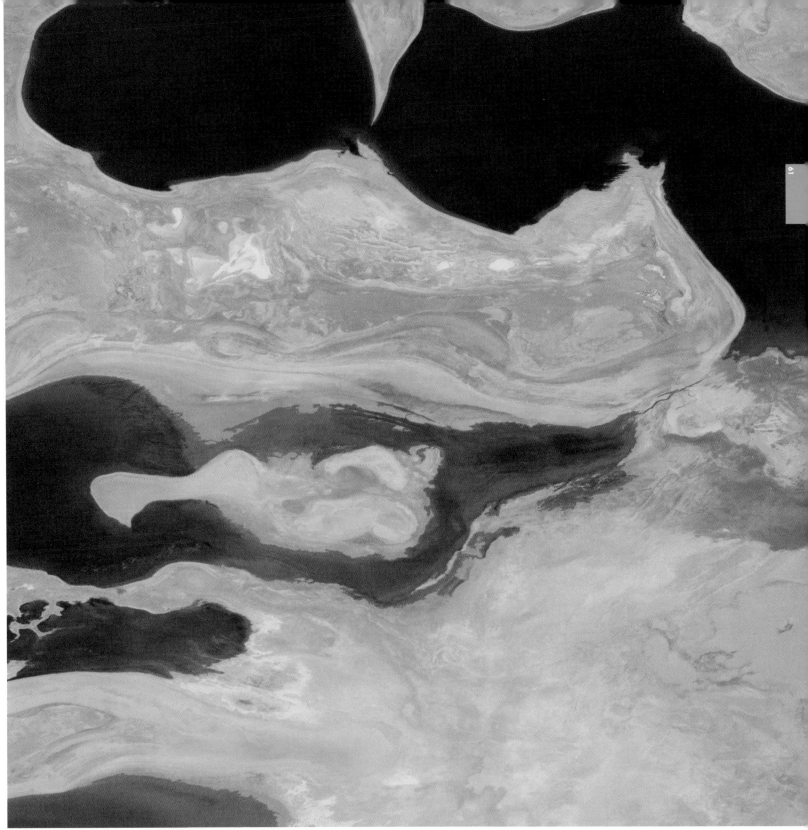

Above: By the time this picture was taken in 2000, the sea was disappearing. The larger southern part, now drying out and separated into a western and eastern half, faces the same bleak future. Complete disappearance could happen in as few as 15 years. The southern Aral Sea has been deemed beyond salvaging, and a restoration effort by the Kazakh government funded by the World Bank will instead focus on the much smaller, but less polluted and saline, northern sea. A permanent wall will be built between the two portions of the sea, sealing the southern half's fate. The northern Small Aral Sea will be allowed to refill from the inflow of the Syrdar'ya river, and though it is never expected to regain its former extent, planners think that it will refill enough to support a robust fishery again. It should also help to stabilise the continental climate, increasing rainfall, smoothing out winter-summer temperature extremes, and suppressing dust storms.

62

Man-made global warming has a short history and a long future. The natural balance of gases in the atmosphere began to change with the industrial revolution, but in the last 50 years, as the whole world has begun to demand a higher standard of living, greenhouse gases in the air have been increasing at an ever faster pace.

It is already clear that the effects of the extra pollutants mankind has released over the last 200 years will be felt by future generations for at least 1,000 years, probably 10,000. As long as fossil fuels are burned at the current pace, never mind the faster speed that most predict, the processes already begun will continue to accelerate, and change the planet for ever. The comprehension of what man is doing to the planet is relatively new. The French mathematician and scientist Jean Baptiste Fourier was the first to realise that the temperature of the earth was controlled by the gases in the atmosphere. In 1827 he observed that certain gases trapped heat. Unusually for a scientist he was able to describe a complex theory so simply that everyone could understand it. He said the atmosphere was like the glass in a greenhouse. It let the sun's rays in and therefore the warmth, but certain gases, notably carbon dioxide, provided a barrier which prevented the heat escaping again. He even coined the phrase "greenhouse effect".

Not that the greenhouse effect is necessarily a problem, in fact it is essential for life on earth. Without this mix of gases which control the heat at the earth's surface conditions for mankind would be unpleasant, not to say impossible. The average temperature has been around 15°C for the last 10,000 years, regulated by the heat-trapping natural gases in the atmosphere, mainly water vapour. Without them the temperature would be well below freezing, about minus 15°C, soon snuffing out most life.

Fourier's theory was developed by a British scientist, John Tyndall, in 1860. He measured the absorption of heat by carbon dioxide and water vapour and suggested, correctly, that a reduction in the amount of carbon dioxide in the atmosphere might have led to a lessening of the

greenhouse effect and the ice ages. Another 36 years passed before a Swedish chemist, Svante Arrhenius, considered the opposite possibility. What would be the result of significant increases in carbon dioxide in the atmosphere? He believed doubling the concentration of carbon dioxide would increase the temperature across the globe by 5°C to 6°C. Although modern scientists would say it was for the wrong reasons, it is remarkable that despite all the progress in science in more than a century, and the immense complexity of the climate system, these calculations are very close to what scientists now estimate will happen — within the lifetime of some people now being born.

Although carbon dioxide amounts were already rising it was not until 1938 that G S Callender, a British meteorologist, tried to raise the alarm. He attempted to persuade the august but sceptical Royal Society in London that global warming was already taking place. He had gathered information from 200 weather stations across the world and demonstrated that in the previous 50 years temperatures had increased. He was not taken seriously.

We know now that Callender was very much on the right track, though probably only a small part of the warming in his day was man-made. It was not until the second world war had come and gone and science was having a special year, the International Geophysical Year of 1957, that the story moves on. The world's scientists were co-operating to study the earth's natural systems, and America comes into the story. Two scientists from the Scripps Institute of Oceanography in California warned that the human race was carrying out a huge experiment with the atmosphere and therefore the future of the entire planet. Roger Reville and

Right, top: A Kazakh villager carries a bucket of water from a well on what was once the bed of the Aral Sea, outside the village of Karateren, southwestern Kazakhstan in April 2005. Dried salt containing pesticides from cotton crops blows across the desert plains, threatening the health of local people. The sea has retreated 150 km (100 miles). One of the rivers that used to feed the Aral, which runs through Uzbekistan, still has all the water removed for irrigation before it can replenish the sea. The other, the Syrdar'ya, is being allowed to flow to the Aral by neighbouring Kazakhstan to try to restore this area to a sea.

Right, bottom: Children run past ruined ships abandoned in sand that once formed the bed of the Aral Sea near the village of Zhalanash, in southwestern Kazakhstan. One of the rivers that fed to sea, the Amudar'ya, once known as the Oxus, was as large as the Nile, but no longer reaches the sea, causing what has frequently been called the single greatest ecological disaster caused wilfully by man. The once healthy people who lived on a diet of fish from the lake and had fertile fields and gardens have serious health problems and there is a high instance of birth defects.

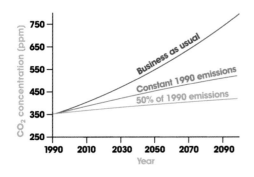

Three possible trends in carbon dioxide concentrations in the atmosphere depending on how the human race combats climate change. According to climate scientists "business as usual" will lead to catastrophe. Their recommendation is a 60% cut from 1990 emissions.

Source: IPCC/Hadley Centre for Climate Prediction and Research.

Hans Suess thought that the build-up of carbon dioxide in the atmosphere could be dangerous.

The warnings of these two men were at last taken seriously, in fact ever since the United States has been in the forefront of investigations into the potential effects of climate change, and actual measurements of what is happening. As a result of these fears routine measurements of carbon dioxide in the atmosphere were begun at one of the most remote places on the planet, the observatory on Mauna Loa, in Hawaii. At 3,300 metres (11,000 ft) above the sea and far away from industrial centres the air is constantly measured to give scientists the world over a record of the composition of the atmosphere. While it always shows a summer and winter fluctuation in carbon dioxide levels, which ties in with the seasons and the absorption of the gas by plants, the trend has been constantly upwards year on year for half a century. The speed of increase is getting faster and faster with each decade. The rising measurements reflect the ever increasing burning of fossil fuels and the greater destruction of forests.

In view of more recent history it is richly ironic that it was the United States scientific community and its politicians that first demanded action to prevent the threat of climate change. This thirst for scientific knowledge and the politicians' acceptance of the need to respond to the grim warnings which began in the 1950s did not change until the late 1980s and 1990s. It was then that the US politicians began to shrink from even appearing to threaten the comfortable lifestyles of their electorate.

The potential inconvenience to voters of tackling climate change was not an issue before then. This was partly because in the 1960s and

70s there had been a relatively cold period in our recent climate, and the prospects of rapid global warming seemed remote. There had even been talk in the 1970s of the return of the ice age, an episode still frequently quoted by global warming sceptics, in an attempt to discredit current climate science.

While nothing adverse appeared to be happening to the weather, despite the fact that measurements showed carbon dioxide continuing to build up in the atmosphere, scientists continued to be concerned. The World Meteorological Organisation, now better known by its initials WMO, sponsored a conference on long term climate fluctuations at the University of East Anglia at Norwich in 1975. This is still a centre of excellence in climate science. The US National Academy of Sciences was also not letting the matter rest and in the same year produced a disturbing report Understanding Climate Change: A Programme for Action. The report was concerned about the probable warming effect on the earth of the emissions from heavy industry. Two years later in a second report the academy went further and warned that the implications of projected climate change "warrant prompt action".

Partly as a result the first World Climate Conference took place in Geneva in 1979. Scientists noted the increased carbon dioxide in the atmosphere and put it down mainly to increased industrialisation, raising for the first time the loss of forests as a cause. The cutting down and burning of forests released stored carbon in the wood back into the atmosphere, the scientists said. A year later a follow-up conference estimated that a doubling of carbon dioxide levels in the atmosphere would cause global warming of between 1.5°C and 4.5°C, an underestimate by current thinking.

Right: The graceful domes of an orthodox church in St Petersburg seen against the smoke stacks and sunset of a winter sky. The picture taken in 2005 was to illustrate the challenge the world faces in tackling global warming. Russia, after months of prevarication, caused when its scientists claimed that global warming might be good for such a cold country, finally ratified the Kyoto Protocol in November 2004, enabling it to come into legal force on February 16th, 2005.

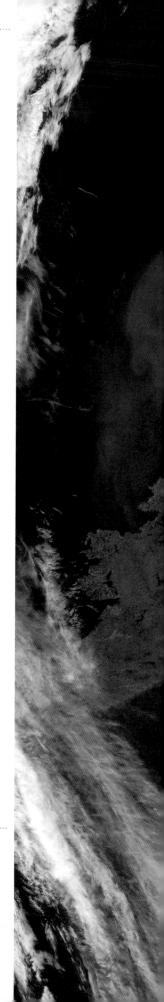

Further meetings in 1985 and 1987 reinforced these findings and suggested an international treaty to cut back on the rate of greenhouse gases in order to reduce the dangers of excess warming. The following year the WMO and the United Nations Environment Programme set up the Intergovernmental Panel on Climate Change, the IPCC, to assess scientific information and formulate response strategies. In June 1988 another significant event occurred. James Hansen, from the National Aeronautics and Space Administration (NASA), testified live on television that he was 99% certain that the warming in the 1980s was not a chance event but linked to climate change. He said: "It is time to stop waffling so much and say the evidence is pretty strong that the greenhouse effect is here." He was speaking at the same time that the United States was suffering a catastrophic drought in the American mid-west which he appeared to be blaming on global warming. At the time his comments created a sensation and were reported round the world.

In the same month in Toronto 48 countries took part in a conference called The Changing Atmosphere: Implications for Global Security. The conference called for a 20% reduction in global carbon dioxide emissions from 1988 levels by 2005 with the eventual aim of a 50% reduction. The conference even talked of a carbon tax to try to reach these aims. These conference targets to reduce greenhouse gases seem in tune with current scientific thinking of what needs to be done. What was wildly optimistic, in the light of recent history, is the timeframe by which those present thought these targets could be achieved. It was clear, though, even in 1988 that the threat posed by environmental damage to the planet had reached the political mainstream.

Sir Crispin Tickell, a scientist and senior diplomat, has been a tireless campaigner about the dangers of climate change for more than 20 years. In the 1980s in the middle of a distinguished career in which he was British ambassador to the UN among many other jobs, he was close to the centre of power. He had studied climate change at Harvard and is credited with convincing Margaret Thatcher of the seriousness of environmental issues and particularly of global warming while sitting next to her on an aeroplane. He also helped write a key speech for her on the issue. Margaret Thatcher, then at her highest profile on the world stage, spoke to the Royal Society in September 1988 about the issues of acid rain, damage to the ozone layer, and the threat of global warming. It was a message she later repeated to the United Nations in New York.

So there was an apparent international acceptance of the problem and, as the conference in Toronto illustrated, already a comprehensive understanding of what action was necessary to deal with the threat. But nothing substantial actually happened. From that moment, it seems, the issue of climate change moved from the purely scientific to the political. The IPCC has become the biggest collaborative scientific effort in history but also the centre of an enormous political circus. Every country in the world is entitled to contribute its expertise. Some, particularly oil producers and coal users and exporters who regard the possible curtailment of the use of fossil fuel as a threat to their economies, pack the proceedings with as many climate sceptics as possible. These are the professional contrarians. They are part of a sophisticated and brilliantly successful campaign funded by the fossil fuel lobby to prevent action on climate change, designed to destroy and discredit the

Right: Sometimes mystified motorists in northern Europe notice that their cars have acquired a rust-coloured tinge caused by red dust. This picture taken in April 2003 captures the source of the pollution, minute particles of sand whipped high into the air by powerful desert winds in the Sahara. The dust can discolour the snow on the Scandinavian mountains, reducing its reflective power and allowing it to absorb heat and so melt.

"It is crazy for us to play games with our children's future... We have a heavy obligation because we now know what is happening to our climate and what will be a highly predictable set of outcomes if we continue to pour greenhouse gases into the atmosphere. We also know we have an alternative of even greater prosperity if we apply the new technologies that are now available to us."

Bill Clinton, former US president speaking at the Montreal Climate talks, December 9th, 2005.

science and prevent political action to reduce greenhouse gas emissions. Details are in the next chapter but a contrarian can argue about any point for hours, for example whether the phrases greenhouse effect, climate change and global warming should be used in official documents. The greenhouse effect, being a natural phenomenon, should properly have the word "enhanced" in front of it to make clear the role man has in causing global warming. On the other hand contrarians do not accept that the globe is warming, or at least that man is responsible, so that phrase is out altogether in favour of climate change, which means the temperature could go down as well as up. Journalists scoff at such pedantry but contrarians have held up whole conferences for hours with such empty discussions.

It is not that scientists, in spite of these arguments and obstructions, ever stopped looking at the real problems. The creation of the IPCC and its subsequent series of reports on the causes and the effects of climate change have been detailed and comprehensive. A vast body of new science on the climate has been created, nearly all of it reinforcing the original view that the world is facing a crisis.

As it has become clear that a rise in sea level is already inevitable and temperature rises are already built into the system, increasing emphasis has been placed on the need to adapt. It is already accepted then that climate change is unstoppable. Estimates of the economic costs of doing nothing about cutting emissions have been increasingly alarming. Perhaps the most important single conclusion of the scientists is that man is clearly responsible for the problem. What they are saying loud and clear is that only political action by world leaders can hope to stop climate change

becoming a catastrophic disaster for both the natural world and the human race. Sadly some scientists regard that as their job done but more should take the example of Sir Crispin Tickell, and do their civic duty and reinforce the message to politicians and the public as often as possible. However, even without that extra push to public opinion, there can be no doubt that the message has been conveyed frequently enough to the world's political leaders for the crisis to be unmistakable.

In fact it is quite clear that in some senses the message has sunk in, because while it appears that despite these unmistakable warnings little has been done it is not because politicians have stopped talking about the problem. There has been an unprecedented series of high level meetings on the environment at which climate change has usually been top of the agenda, and sometimes the only topic. A list of them would run into hundreds, many of them appearing to have achieved very little. But international diplomacy is like that, incremental rather than dazzling breakthroughs. There have been some important milestones along the way, all providing an international framework so that the world leaders could take action on global warming if they could summon the political will. General concern about the degrading of the earth's environment was the impetus behind the Earth Summit of 1992 in Rio de Janeiro. A centrepiece for the summit, then the largest the world had seen, was the United Nations Framework Convention on Climate Change. It is a remarkably short and clear document, rarely quoted, but in the light of subsequent events, worth quoting. The "objective" of the convention is but a single paragraph. It "is to achieve ...stabilisation of greenhouse gas concentrations in the atmosphere at a level that would prevent dangerous anthropogenic (man-made)

Left, top left: Another "historic moment" in the climate negotiations. Here the Japanese environment minister, Hiroshi Oki, is seen on two screens as he addresses the final plenary session of the United Nations global warming conference in Kyoto on December 11th, 1997. This was where the protocol aimed at heading off a potentially catastrophic warming of the earth, and now known by the city's name, was formally adopted after 11 days of tense debate.

Top right: After the US walked out of the climate talks yet again, this time in December 2005, Bill Clinton, the former US president, rose to speak, calling on his successor George W Bush to come back to the negotiating table to safeguard the future of the planet.

Bottom left: While Bill Clinton uses his persuasive powers inside the conference hall thousands of people march outside through the streets of Montreal in December 2005 as part of a worldwide day of protest against global warming.

Bottom right: A Greenpeace member protesting at the Bush administration's repudiation of the Kyoto Protocol at the climate talks in Bonn in July 2001. It was the meeting at which the rest of the world decided to proceed with the legally binding agreement and ignore American attempts to destroy the treaty.

interference with the climate system. Such a level should be achieved within a time-frame sufficient to allow ecosystems to adapt naturally to climate change, to ensure that food production is not threatened and to enable economic development to proceed in a sustainable manner."

The first key principle on the next page says: "The parties should protect the climate system for the benefit of present and future generations of humankind." It goes on to say that the developed countries, that is those which had already grown rich on polluting the atmosphere, "should take a lead in combating climate change and the adverse effects thereof". Many nations signed up to it immediately. As the key principles go on to explain, the convention puts the onus on those who had signed and ratified the agreement in their parliaments to "adopt policies and measures" to achieve a safe climate. Countries agreed as a first step that by the year 2000 they would get carbon emissions back to 1990 levels, a promise all the politicians apparently forgot the moment they signed up to it. In any event they did nothing to make good their pledges. The lesson learned here was that countries were obliged in the treaty to take measures to tackle climate change but in effect the convention was rendered toothless because the target was not legally binding. It took more than 10 years to rectify this problem with the adoption of the Kyoto Protocol. The 1992 summit was important for other reasons, both for the agreements it reached, and those it failed to. Another success was the Biodiversity Convention, designed to prevent the continued extinction of animals and plants. It was also agreed and signed, although the conspicuous reluctance of the United States to take part because it might cost too much money and affect American interests was an ominous

Left: Deliberately lit forest fires visible for miles in Para state, Brazil are designed to clear the land for soya bean production. Although the Brazilian government is attempting to stop forest clearances they are largely ignored by large scale farmers as this 2003 picture shows. Producers are anxious to cash in on the profits available for soya exports to food processors in the United States and Europe. Many packaged foods contain soya.

indication of what was to come. A forest convention, billed by Europe and the United States as a major object of the conference, never materialised. The developing world refused to accept any interference with the sovereign rights to control their own natural resources. Fourteen years later a forest convention is no nearer fruition. One of the surprises of the conference was the move by developing countries to create a convention to combat the spread of deserts. Although it was barely of interest to the developed world, which apart from the US and Australia does not have major deserts, the convention moved forward quickly. The subsequent programme of international co-operation and measures to push back desert frontiers has been the least reported but one of the most successful outcomes of the Rio summit. There was also an extremely long and largely unread document called Agenda 21, which was a blueprint for the 21st century, designed to get the planet on course for a sustainable future. Many local, rural and city governments have subsequently taken its provisions seriously and made a huge difference to the quality of local life. A far greater number have ignored it.

The whole summit ended in a mood of optimism. At last there was a feeling that there would be international action to preserve the environment. Even though people were aware that there were tensions between the developing countries and the rich nations, everyone was heading in the right direction. There are a number of important themes that emerged which are as influential 14 years later. The developing countries' point, still argued strongly, is that the rich countries grew rich by wrecking the environment, and since they had caused the problem they should bear the brunt of fixing it. Meanwhile the developing countries'

Left: The tribal lands of Yanomami Indians of the Roraima region of the Amazon are supposed to be protected areas, but as this picture shows logging continues anyway. The loggers are followed by farmers who clear the remaining forest. Although the Amazon is still an area of tropical forest larger than Europe it is being eaten away at an ever growing rate. The Amazon is known as the lungs of the world because of the amount of carbon dioxide the forest absorbs. This free service in helping to keep the climate stable is being lost as the trees are destroyed.

"The arguments about the science of climate change are now over as far as we are concerned. There is so much accumulated evidence about the link between human activity and climate change the only issue left is how to tackle the problem."

Elliott Morley, then the UK's environment minister, 2005.

priority was, and is, to continue developing. This was the principal reason for the foundering of the forest convention. As has already been said, developing countries regarded their forests as their own natural resource and if they wanted to cut them down, either for cash or for development, it was their sovereign right to do so. Thinking on this point is at last beginning to change, bringing new hope both to poor countries and to the remaining forests, but more of that later. In 1992 they were pointing out that developed countries had long ago cut down their own forests for agriculture and for building materials for boats and houses, so were in no position to criticise poorer countries for doing the same. One significant point about all these agreements and negotiations, successful or otherwise, was how they were inter-related. There was the overall theme of how to develop while at the same time avoiding the destruction of the environment. But just as important, as we see in the science of climate change, all these environmental matters are related. How can climate change be considered without also looking at the loss of forests, the spread of deserts and the disappearance of so many plants and animals?

The optimism of the Earth Summit proved to be a false dawn, partly because the world was in the middle of a mini-recession where jobs were being lost and western economies stagnating. The environment took second place to the need to stimulate growth. There is also a time lag built into international treaties, which involves waiting for enough governments to pass legislation through their parliaments to bring any convention to fruition.

In the case of the Climate Change Convention this was remarkably swift compared with other attempts at international regulation, which have

sometimes languished for more than a decade before reaching a critical mass with enough participating nations prepared to make parliamentary time to bring them into force. In the case of the climate convention it took only three years before 50 countries of the 150 which had signed up to it in Rio to ratify, opening the door for the first Conference of the Parties (or COP 1) in Berlin in 1995. Under the convention rules, once started, these COPs are held every year. They now stretch over two weeks from the end of November into the first week of December, usually with a full attendance list of politicians, indicating their importance. Mid-way through the year in June a lower level meeting of civil servants is held to assess progress from the last meeting and prepare the ground for the next. The landmark meeting in Montreal in December 2005 was COP 11, but more of that later.

The result of the first COP, the Berlin meeting, was a mandate to seek a legally binding agreement within two years to cut greenhouse gas emissions. It was already acknowledged that the voluntary agreement, reached in Rio but ignored, was never going to deliver a safer climate. One of the features of the Berlin meeting, and previous political gatherings about climate change, was the emotional but completely reasonable appeal of the Alliance of Small Island States (AOSIS). This was the first major grouping of countries concerned about climate change. They were the first to realise that the theoretical threat of sea level rise, and storm surges, had for them become a present danger. Many of these states, like the Maldives and the Marshall Islands, are based on vulnerable coral islands. Others, like the Seychelles or the Caribbean islands, have some high land but are already losing beaches on which their main foreign currency earner,

Right: Thunderstorms are common in the tropics and particularly over the forests of the Amazon, and bring large quantities of rain, which gives the area its character. A NASA-funded study has shown that tiny airborne particles of pollution caused by forest fires modify developing thunderclouds by increasing the quantity of water droplets and reducing their size at the same time. Researchers believe that rainfall over the forested areas is reduced by 20%, enough to cause considerable changes in the tree cover below. The immense forest is home to thousands of plant and animal species, many of which are still to be recorded. One of the top predators is the elusive jaguar at home in the densest forest. Continued logging threatens its future and that of many other species. Thousands of plant and tree species that occur nowhere else on the planet will also be lost forever. Many have medicinal properties known to the indigenous population but not yet investigated by modern scientists.

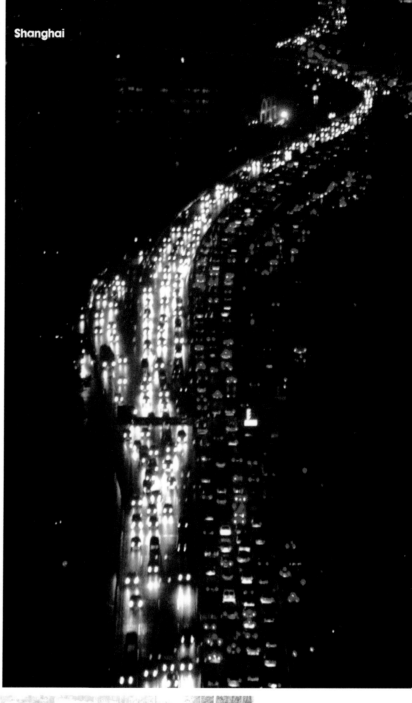

Shanghai

Los Angeles

São Paolo

Dhaka

tourism, depends. For many of these 36 AOSIS states, increasing storms and weather-related disasters mean at a minimum economic instability, and for others, in the worst case, sea level rise means the end of their existence as nations. Whole populations will have to leave their homelands for ever and seek sanctuary elsewhere.

At many of the international conferences, where world leaders meet, this impending crisis for these most vulnerable countries has been stridently and often emotionally put by various prime ministers and presidents of the countries involved. They have been politely listened to, and loudly applauded, but in terms of action, largely ignored. As early as 1990, when AOSIS was originally formed, the grouping had clout in the United Nations system because of the one nation, one vote principle. Aided by a handful of young and enthusiastic international lawyers the group helped to draft the original Climate Change Convention and raised the issue of the inability of these vulnerable low-lying nations to get flood insurance, and the crucial question of what would happen to their peoples when sea levels rose. They were an important counterbalance to the obstructions of the fossil fuel lobby and oil-producing countries like Saudi Arabia.

In Berlin, the AOSIS demand at the meeting for a 20% cut in carbon dioxide emissions by industrialised countries by 2005 was specifically included in the mandate. The culprits, that is the industrialised countries, and the need for the governments of those states responsible to sort out the mess, were again clearly spelt out.

Two years later, as the so-called Berlin Mandate had demanded, a third and as it turned out

highly dramatic meeting took place in Kyoto. There had been many tortuous preparations but the 10,000 people who gathered in Japan were in for a gruelling time. Through all the jargon, horse-trading and lobbying from all sides the deal was about how much each of the industrialised nations would agree to reduce their carbon dioxide emissions. Everyone was prepared to accept that every country had a different set of circumstances. Some were at an early, some a late stage of development, and some had special industries that used a lot of fossil fuels. Every country seemed to have a case that it should have a less demanding target than its neighbour. Most pressure of all was on the United States, by far the world's biggest polluter, to take responsibility for the mess. The figures speak for themselves. A country with about 4% of the world's population was and still is responsible on its own for 25% of the carbon dioxide emissions. With Bill Clinton claiming to be a green president, and his vice-president Al Gore already having written a book on the dangers of climate change, it was seen as politically impossible for the US not to do a deal. In Kyoto, however, hope seemed to fade. After two weeks of tough talking, with the conference due to end in two days and no deal in sight, it was decided to continue night and day in continuous session until an agreement could be reached. Forty-eight hours later the conference was still in session, people were literally dropping asleep in their chairs, television crews and reporters slept in corridors.

Long after the conference was supposed to have finished a deal was done. Thirty-six industrialised countries, including crucially the EU, Japan and the US, had agreed to legally binding targets to reduce greenhouse gases from 1990 levels by 2008-12. The rather odd

Left: Car ownership is increasing all over the world and there are few cities anywhere that do not have traffic jams. As this packed freeway in Los Angeles shows, however many roads you build there is never enough space for all the cars that want to use it. In the booming city of Shanghai in September 2005 the lights on the Yanan elevated road show that China is fast catching the developed world in car ownership. Makers from all over the world are trying to break into this number one growth market, but it is not clear whether there is enough room for all the vehicles. In Dhaka, the capital of Bangladesh, one of the poorest countries in the world, where only a very small percentage of the population can afford a car, total gridlock and smog are already part of daily life. Teeming São Paolo in Brazil cannot keep up with its booming traffic either.

"An increase of two or three degrees wouldn't be so bad for a northern country like Russia. We could spend less on fur coats, and the grain harvest would go up."

Vladimir Putin, Russian president,
October 2003.

four-year period was the subject of much negotiation because some countries thought they could do it at the beginning and others not until the end. No targets were set for developing countries. Industrialised countries would reduce their collective emissions of greenhouse gases by 5.2% compared to the year 1990. That may not seem much but the United Nations Environment Programme noted that it was a 29% cut on what would have been expected by 2010 without the protocol.

The bottom line for individual countries reflected the horse-trading. The US had agreed to a cut of 7%, Japan 6% and the European Union 8%. In fact the EU figure hid a complex deal. Some countries in the 15 thought they could do considerably better, notably Denmark, which had a large renewables programme, and a unified Germany, which had already closed down a lot of energy intensive industries in the former East Germany. Both were prepared to take a 21% cut. The United Kingdom, which had closed a lot of coal mines, and was transferring energy production to more efficient gas fired power stations, promised 12.5% cuts. This enabled the less developed countries of south- ern Europe increases, with Greece allowed 25% and Portugal 27%. Overall this allowed an impressive 8% cut for the EU without reducing growth. In the event this is proving a tough legally binding target to meet, with considerable doubt that it will be achieved. Most of the former Soviet bloc countries, apart from Russia, followed the EU with an 8% cut, partly through solidarity, but mainly because they too had lost inefficient industries with the collapse of the communist regimes. Switzerland, which had no reason, apart from desperation about the effect of climate change on the Alps, opted for a difficult 8% cut because it saw the size of the threat to its tourist industry.

Russia, with its vast forests, and collapsed industries, was allowed to get away with holding emissions at 1990 levels. This meant that Russia would easily meet its targets and this was interpreted by the environmental movement as a fiddle in the making. Under the terms of the Kyoto agreement it was possible for countries that exceeded their targets to sell the tonnes of carbon saved to those that had failed. This system is designed to reward financially those that have done well and allow those which have failed to buy their way out of a problem. It was felt Russia was getting a free ride and it would be a way for the wealthy United States to purchase "hot air" to avoid having to take action at home. In the event Russia is still well placed to exceed its target and cash in on the surplus carbon, but its main potential customer, the US, has vanished, having repudiated Kyoto. Judging by the failure of many countries to take enough action to curb emissions and so meet their targets Russia will still not be short of customers by 2012. Some countries were allowed rises, as much as 10% in the case of Iceland. Australia, with its large fossil fuel industries, insisted on plus 8%, a curious anomaly as it turns out. Australia, having repudiated Kyoto, now seems likely to reach its target through industrial changes that would have happened anyway. It shows that predicting the future is an inexact science.

At the signing of the Kyoto deal the world heaved a sigh of relief — at last something was being done. In fact since those heady days in Kyoto international negotiations on climate have been long drawn out, painful and essentially depressing, although not without excitement. The high spot in drama was when the newly elected President George W Bush repudiated the Kyoto agreement exactly three years after it had been agreed. This was partly for ideological

reasons, but also because the US economy had grown so much since 1990 that America could not hope to reach its targets. As President Clinton pointed out in his hard-hitting speech at Montreal during a crucial moment in the 2005 climate talks, he had in his last years as president tried and failed to bring in legislation that would reduce America's emissions.

It had become clear to the Bush White House that the US's production of greenhouse gases had increased so much during the prosperous years of the late 1990s it was already impossible to reach the 7% reduction target, even if there was an ounce of political will to do so. But even before the election it was clear the oil man, George W Bush, was the preferred choice of the fossil fuel lobby, which heavily financed his campaign. The speed with which he repudiated the protocol on taking office was the only surprise.

But just when many thought all was lost, and perhaps uniquely in international relations, the rest of the world decided to ignore the world's richest and most powerful nation and continue with Kyoto. Six months later, at what would have otherwise been a routine meeting of civil servant advisors the politicians turned up in force in Bonn, the headquarters of the climate change convention secretariat. There was another exciting moment in the history of the convention, when the remaining industrialised countries which had targets refused to let the agreement die. In effect, at least as far as climate change was concerned, they made the United States an international pariah.

In the following years the US administration has done its best to undermine Kyoto. The convention's main champions have been the European Union and Japan. Apart from other motives Japan has an emotional attachment to the treaty. Since it was signed in a Japanese city, in addition to any other reason, it is a matter of honour and pride for the nation to make it work.

In the five years since the high point in Bonn when Kyoto was rescued the political process has stalled. This was partly an inevitable consequence of the slowness of the legal process of bringing it into force. By the end of 2003 120 countries had ratified it, which under the terms of the protocol was well in excess of the 55 needed, but there was another provision more difficult to fulfil. This was to do with the emission targets that the protocol had set. Among those ratifying the convention there had to be enough of the industrialised countries with targets to reduce emissions to represent 55% of the total greenhouse gases released by the industrialised world. Now that the United States, with 25% of the emissions, was out of it, this meant all the other big polluters had to ratify to bring the protocol into force. At the end of 2003 the total of emissions of the ratifiers was only 44%, which included the EU, eastern Europe and Japan. Russia's 17% was needed to reach the necessary total, and the Kremlin was playing hard to get. President Vladimir Putin had more than once said Russia would ratify but some influential Russians from the Academy of Sciences resisted the idea. A bit like some of the climate sceptics in the United States they poured doubts on the science but behind the scenes the motive appeared to be a throwback to the cold war. There was some evidence that climate change would be an advantage to Russia, allowing wheat and other crops to grow further north than now. Changes in rainfall patterns would also benefit some areas, while climate change was clearly more detrimental to the old enemy, the United States. Disadvantages

Right: The US president, George W Bush, hands a pen to the then German chancellor, Gerhard Schröder, as they sign a communiqué at the G8 Summit in Gleneagles, Scotland, July 8th, 2005. Behind them from left are the French president, Jacques Chirac, the Japanese prime minister, Junichiro Koizumi, the British prime minister, Tony Blair, the then Italian prime minister, Silvio Berlusconi and the Russian president, Vladimir Putin. The world leaders announced a $50 billion boost in aid and debt relief, much of it to Africa. There was also a declaration on climate change, the first time George W Bush had acknowledged it was a man-made threat, but there was little action promised.

to Russia, including the massive melting of the permafrost and the loss of some forests, were discounted against these advantages. The argument in Moscow was that Russia should not be in too much of a hurry to slow down climate change.

In the end, having used Kyoto as a bargaining chip to extract some backing for trade deals from the European Union, President Putin kept his promise and the Russian parliament ratified the agreement. Kyoto came into force on February 16th, 2005, to much international rejoicing. (The word historic cropped up a lot). Outside the conference halls the struggle to reach critical mass for ratification had grabbed a lot of attention. There was a lot less interest in the equally important squabbling between the parties over the details of the convention. The obvious way to cut emissions is to burn less fossil fuel but there were details of various schemes to negotiate, including those to introduce clean technologies to other countries in exchange for carbon credits. One of the most difficult discussions, which went on for years and will continue, is about the rules of how various greenhouse gases could be reduced by growing or replanting forests. There is a debate about how much carbon dioxide is emitted from the ground when it is disturbed to plant new trees, and how much carbon is absorbed by the saplings. In the longer term how much can these totals count against any carbon dioxide emission targets when forests can be burned down, cut down or subsequently harvested? The United States and its one ally Australia, while remaining outside Kyoto, continued to do everything in their power to undermine the political process, including raising difficulties in all these discussions. As time passed the next phase of what to do about climate change beyond the expiry of the Kyoto Protocol in 2012

began to loom. Efforts were made by Tony Blair, the British prime minister, among others to woo the White House, because it is clear that without the world's largest polluter in the process attempts to save the climate will fail.

At the same time the rapidly industrialising developing world, notably China, India, Brazil, South Africa and Mexico, were increasing greenhouse gas emissions at an alarming rate, and were not bound by targets under Kyoto. Their involvement in the process must be increased so that they can become more active participants in reducing their own emissions as soon as possible. With Britain's turn as president of the G8 group of industrialised countries coming up in 2005, Tony Blair made a bold move. He announced that he was making the end of poverty in Africa and climate change his twin priorities for the British presidency. As part of this the leaders of the big five developing nations named above were invited to the G8 meeting at Gleneagles in Scotland in July 2005 specifically to discuss climate change. They accepted and progress was made on debt relief for the world's poorest nations but critics said no concrete progress was made on climate. This was slightly uncharitable since President Bush did make some statements about climate change being a threat. This had long been acknowledged almost everywhere outside the White House, but nonetheless for the US it was progress of a sort.

Barely a month later, however, when news began to leak out of a deal which the Bush administration had done behind Blair's back to share new technology with China, India, Japan, South Korean and Australia, suspicions about US motives over climate were rekindled. This agreement was seen as another attempt to undermine Kyoto, although all the countries

Left: Traffic and pedestrians hustle through Shanghai's Nanjing Street, famous throughout Asia for its shopping. China, and particularly this bustling port city, has embraced consumerism and the western way of life. With the large port and commercial area only two metres above sea level the city is in danger from sea level rise and is already spending huge sums on building its defences against tidal surges caused by ever worsening typhoons in the region.

"Unfortunately, some parties, by not abiding by their commitments, have put the credibility of the Kyoto Protocol in question."

Rafiq Ahmed Khan, high commissioner of Bangladesh, December 2005.

involved have subsequently denied that. Japan's involvement meant that at least the government in Tokyo was sincere in believing it was an extra agreement rather than a replacement, and at Montreal, China's statement that Kyoto was the only credible political agreement to tackle climate change has removed any lingering doubt about its status. If anything, while progress in reducing emissions has been painfully slow, the political rhetoric and apparent alarm at the imminent threat has increased dramatically. This is in line with the scientific findings of the speed of the changes the world is now facing. The phrase first used by Sir David King, the UK's chief scientific advisor that "climate change is a greater threat to the world than terrorism" at the beginning of 2005 has been repeated many times since by politicians, including Tony Blair, to the point where it is already a cliché. Yet 18 years after the Toronto conference the target of a 20% reduction of carbon dioxide levels of 1988 has not yet been attempted. In fact total world emissions have gone up at an ever faster rate since then, and on present trends that seems likely to continue for decades.

It would be churlish to say there has been no progress. Considering the pace of other international agreements of the past there has been a whole series of advances in terms of international treaties and agreements. The level of education on the subject at all levels of society has come on at great speed, not least because the changes are already noticeable and widely reported. It would be fair to say the general public across the world is now aware how the climate is changing. It is hard to go anywhere in the world where the older generation will not happily recount changes in their local weather.

In Europe, particularly Britain, where talking about the weather is often the conventional start of any conversation, any unusual variation is put down to global warming. Hosts of people now monitor the first frog spawn of spring, when hedgerow birds begin to nest, and when leaves sprout and fall, all in an effort to keep track of climate change. Hardly a day goes by without details of new science related to global warming being widely reported. Books are published across the world explaining how to green your lifestyle. Anything from a dozen to a hundred ways to save the planet are urged on readers of newspapers and magazines.

Yet the political effort to do something about the causes of climate change is at best slow and ponderous. At worst it is fiddling while the earth burns, and an awful abdication of the responsibility of leadership.

At the end of 2005 eyes then turned to the COP 11 meeting in Montreal. This is where the world's experts in the politics of climate change met to explore the future. It was also the first meeting of the parties to the Kyoto Protocol, the so-called MOP 1. What happened there, at least partly, determines the future of the world as we know it. This was hard to discern from the newspaper and television coverage. The meeting ended with some expressing delight and hope. Others were not so sure. Montreal, and where we go from here, is examined later.

Left: A motorcyclist turns back due to intense heat as they pass through haze near burnt peat land in Indonesia's Riau province. These fires in August 2005 were typical of those which beset the region as drought and development encroach on forests and peat bogs. In some years the dense smog has choked areas over hundreds of square miles.

The Kyoto Protocol

The Kyoto Protocol was agreed in 1997 and came into force on February 16th, 2005. Originally 36 industrial countries signed up to targets to reduce or control emissions. Two, the United States and Australia, later dropped out. The targets of the 15 member states of the European Union of the time were set jointly so that some EU countries could have higher targets than others. This would allow newly joined southern European states and Ireland, which were regarded as underdeveloped, leeway to increase emissions. To make up the difference Germany, Denmark and the United Kingdom all agreed they could reduce their emissions significantly more than 8% to make up the difference. Most eastern European countries decided to have the same target as the EU, partly because most of them wanted to join the union.

The targets cover emissions of the six main greenhouse gases:
Carbon dioxide (CO_2);
Methane (CH_4);
Nitrous oxide (N_2O);
Hydrofluorocarbons (IIFCs);
Perfluorocarbons (PFCs);
Sulphur hexafluoride (SF_6)

In the jargon these are called "the basket of greenhouse gases". The last three are more correctly families of complex gases, nearly all of them produced by industrial processes. They can mostly be captured and destroyed, re-used or recycled to avoid adding to global warming.

The agreement was to reduce or control emissions of six greenhouse gases. The 1990 level of emissions was used as a base line, and the dates that the target had to be reached was any time between 2008 and 2012.

-8%	The 15 European Union countries, plus Bulgaria, Czech Republic, Estonia, Latvia, Liechtenstein, Lithuania, Romania, Slovakia, Slovenia and Switzerland all agreed to reduce emissions by 8% by 2008/12.
-7%	The United States agreed -7% but later dropped out.
-6%	Canada, Hungary, Japan and Poland.
-5%	Croatia.
0%	New Zealand, Russian Federation and Ukraine.
+1%	Norway.
+8%	Australia (later dropped out).
+10%	Iceland.

Kazakhstan, which was not an original participant in the Kyoto process, has said it wants to join and have a target, but so far one has not been specified. Monaco subsequently ratified in May 2006.

Carbon credits

As well as straight reductions of these gases there are a number of credits nations can earn towards their targets. These include taking action in other countries to reduce their emissions. The idea is that since the atmosphere is a shared resource then a country taking action anywhere to curb emissions is still serving the common good. There are a variety of ways of doing this, all carefully policed by various convention bodies. In addition there is a system of carbon trading where a country which has exceeded its target can sell the tonnes of carbon saved to another country which has failed to do so. This system has been adopted internally inside the EU and internationally. Some believe it will allow Russia and some eastern European countries to profit at the expense of countries that are failing to reach their targets. Others accept that even if that is so, carbon trading allows the industrialised countries to reach a collective goal to reduce emissions and establishes a system which will allow more stringent targets to be set later on.

One important point about the convention is that it has no end date; the idea is to continue adding new targets and timetables until a safe climate has been achieved. Each year countries have to report progress towards reaching their targets and how they propose to make up for any deficiencies. There are fears that many countries are not on course to reach targets and in March 2006 a compliance committee was set up under the Kyoto Protocol to address the problem. It can make suggestions and embarrass countries that are failing to keep to their legally binding targets.

Penalties for not reaching the target, apart from diplomatic embarrassment, come after 2012. By that time a second set of commitments up to 2020 should have been negotiated. Those countries that have failed to reach their first target have it added to the second period, plus a 30% penalty for the shortfall.

Mad, Bad or Greedy?

"I have long believed that claims of a consensus that man is causing global warming is the greatest hoax ever perpetrated on the American people."

James Inhofe, Oklahoma Republican senator and chairman of the Senate environment and public works committee, May 2006.

Contrarians, or climate skeptics, as they like to call themselves, spelt with a "k" here since they are mostly Americans, are taken seriously by the media.

They are a handful of people who crop up hundreds of times in press cuttings and on air, casting doubt on the scientific evidence of global warming and its potential dangers. Some are no doubt sincere, and independent in their views. Many of them are paid by fossil fuel interests, and freely admit it, although it seems to do nothing to damage their credibility. Their apparent aim is to spread confusion as to whether climate change is real at all, or if it is happening, to question whether man has any responsibility. Sun spots, a change in the earth's axis, general unexplained natural variability and references to warm periods in the past are all used to rubbish the mainstream science, even though these factors have already been taken into account by mainstream climate scientists. The second part of the message is to say that action on climate change will cost vast sums of money, wreck the American and world economy, damage international trade, and cost millions of people their jobs.

These sceptics are outnumbered 1,000 to one by scientists who usually have better and more relevant qualifications, along with impeccable credentials. These are people who have increasingly serious concerns about the fate of the planet, and humanity, but who stick to properly reviewed scientific data to make their case. They find themselves rubbished on radio and television by people funded to do so by the fossil fuel industry, who claim man-made climate change is unproven. Newspapers, the Wall Street Journal among the worst, allow their journalists to write thousands of words quoting people with shaky qualifications, often in different fields, who produce junk science. This is all in support of their editorial position that global warming is not a problem and should not interfere with the interests of businesses of all kinds, which should be allowed to continue to

pollute as usual. The point is that these sceptics are not operating alone. There is a network of anti-climate science think-tanks, lobby groups and non-government organisations passing themselves off as concerned for the public good. A key to such groups is the names that crop up in their literature, and are quoted with approval by them. Experts like Dr Robert Balling, Dr Fred Singer, Dr Patrick Michaels, Dr Richard S Lindzen, Philip Stott, Gerhard Gerlich, Willie Soon and Sallie Louise Baliunas, who received an award for her "devastating critique of the global warming hoax". One of them, Dr Lindzen, has produced good papers on water vapour and its effect on climate change, which have challenged existing computer models. But, as usual with the sceptic lobby, his name has been mixed in with less credible and less quali-fied propagandists. All of them are climate sceptics. Most derive considerable income as a result. If they are sincere they are closing their eyes to the weight of evidence. Are they gen-uinely mistaken, mad, bad or greedy?

Being against accepted climate science is a good living. Sixty organisations have been identified in America alone as taking money from the fossil fuel industry to rubbish global warming science and to campaign against any action designed to curb the burning of oil and coal. There are more in Europe and Australia, most of them sponsored directly by industry. Exxon, the world's biggest oil company, has since 1998 given over £10 million ($18 million) to groups opposing the Kyoto Protocol. But this is not a new phenomenon. The fossil fuel industry realised early that fears about climate change were a potential threat to its business. The message has altered over the years, and become more subtle, but during that time the key messengers have worked their way to the centre of power.

91

Previous spread: An oil lake lies near a blazing pipeline outside the southern Iraqi city of Basra, March 29th, 2003. This was a scene familiar across the whole of Kuwait a decade earlier when the retreating Iraqi army destroyed the oil wells and let millions of gallons of oil flow into the desert. Then they deliberately set fire to them. In this case the Basra

South oil refinery, which can produce 140,000 barrels per day, was largely undamaged during the American-led invasion and workers said it was ready to go on stream as soon as electricity and crude supplies were resumed.

Left, top: American flags fly in front of General Motors Corporation's headquarters in Detroit, Michigan, July 2005. Big consumer incen-tives were needed to offload gas-guzzling models as oil prices stayed high because of damage inflicted by hurricane Katrina, continu-ing problems in Iraq, and anti-American feeling in central America.

Left, bottom: An aerial overview of the Peabody Energy Black Mesa coal mine taken in 2001 on the Hopi Reservation. The giant coal mining company pumps water from an aquifer beneath Hopi and Navajo lands for a slurry pipeline. The two peoples claim the mining company is sapping their water supply and con-tributing to a drought.

Peabody Coal has been in the forefront of a successful 16-year campaign to pre-vent the United States taking any action to curb green-house gases and cut the use of fossil fuels.

Bizarre though it may seem now, the fossil fuel lobby's first significant campaign was to promote the need for far more carbon dioxide in the atmosphere. Fred Palmer, then head of a coal company called Western Fuels, now subsumed into Peabody Energy, the world's biggest coal producer, conducted a campaign at the beginning of the 1990s to triple the amount of carbon dioxide in the atmosphere. He made a film, The Greening of Planet Earth, which put forward the idea that crop yields would be boosted between 30% and 60% if carbon dioxide was pumped into the atmosphere. The earth would enjoy an eternal summer and world hunger would be abolished. Among those who saw the film just before the Earth Summit in 1992 were President George Bush Sr, his chief of staff John Sununu, and Bush's senior energy secretary James Watkins, who mentioned the movie as a credible source in interviews about climate change. Mr Palmer is now a vice-president at Peabody Energy, a company that has heavily funded the presidential campaigns of George W Bush. The organisation he formed still exists as the Greening Earth Society.

As climate talks continued during the 1990s the funding of sceptic scientists and lobbyists began in earnest. It has long been a feature of United Nations negotiations that companies cannot have a direct voice, but non-government organisations and pressure groups can have accreditation to lobby delegates with their point of view. Journalists were familiar with the sometimes slick photo-opportunities of Greenpeace and earnest youngsters from Friends of the Earth and WWF hovering on the edge of press conferences providing worthy quotes, but were unprepared for the surge of men in suits. One of the first groups to be identified as "the other side of the argument" were the Global Climate Coalition, supported by

the fossil fuel lobby, car manufacturers and other heavy industry of the old energy-intensive type. About 50 companies that thought they might suffer loss of sales if there were restrictions on carbon dioxide emissions paid to belong. Note the name. It is sufficiently misleading to make the casual observer believe they might care about the climate. Their slogan was "Growth in a global environment". In the mid-1990s any of the numerous meetings involving climate science or politics might feature up to two dozen such organisations. They would fetch up anywhere in the world. Each one came with lobbyists looking for delegates to convince of their case, or as it seemed to me, at least confuse everyone as much as possible.

There are now even more. Among them are the Alliance for Climate Strategies, which includes the American Petroleum Institute among its members, and the Cooler Heads Coalition formed "to dispel the myths of global warming by exposing flawed economic, scientific and risk analysis". Misleading names are an important part of the game. The Centre for the Study of Carbon Dioxide and Global Change promotes junk science as an "antidote" to mainstream government sources.

The most famous lobbyist of them all, and the most effective, was a man called Donald Pearlman, known by journalists and environmentalists alike as "deputy dog" for his jowls. He was a lawyer, partner in the law firm of Patton, Boggs and Blow in Washington. He was always hanging about in corridors, never without a cigarette in his hand. For years he never missed a meeting, however insignificant, on either climate science or politics. He was first exposed in 1995 by the German magazine Der Spiegel, which called him the high priest of

Above: This is where the Texas love affair with oil began. Under the drill the noise started as a rumbling, then turned into a deafening roar as thick black liquid exploded 150 feet (50 metres) into the air, tossing heavy sections of pipe around like a handful of matchsticks. The discovery 100 years ago at Spindletop Hill near Beaumont marked the birth of the modern oil industry in Texas. It ushered in an era of prosperity, something the state is keen to hang on to.

"Politicians from other countries pretend that with enough patience and under-standing they will bring the world's largest emitter of carbon along with them to address this unprecedented global threat. They will not. In the US the White House has become the east coast branch of Exxon Mobil and Peabody Coal, and climate change has become the pre-eminent case study of contamination of our politics by money. What is missing from all these discussions is the sense of desperation and helplessness which is shared by all of us who are shaken by the effect of each new impact of our inflamed atmosphere."

Ross Gelbspan, an American author and
Pulitzer prize winner, during the Montreal climate talks,
December 8th, 2005.

Top: The Exxon Corporation chairman, Lee Raymond, (left) shakes hands with the Mobil Corporation chair-man, Lucio Noto, at a press conference in New York on December 1st, 1998. The two discussed Exxon's acquisition of Mobil in an $80.1 billion stock transac-tion to create the world's largest oil company. The new company continued to question the existence of climate change, funding sceptical scientists and pressure groups, while rejecting involvement in the renewables industry.

Above: US President George W Bush delivers the State of the Union address while Vice-President Dick Cheney (left) and House Speaker Dennis Hastert (right) look on in Washington on January 31st, 2006. It was the first time the president had expressed concern about dependence on imported oil and the need to reduce it.

Top: There are thousands of old tankers, frequently crewed by poorly trained seamen from developing countries, delivering crude oil to its markets. Many travel from the Middle East thousands of miles to oil refineries in America, Europe and Japan. Some of the oldest are retired from the oil trade but continue in service for other purposes. This

is one of them, the MV Ulla, carrying toxic ash from three thermal power plants in Spain in an attempt to export it to another country. Greenpeace called on Spain to take the waste back in 2000 but the ship was still at sea four years later and sank outside Iskenderen, Turkey, causing a major pollution problem.

Above: When oil tankers sink their cargo frequently floats on the sea for days, threatening wildlife and holiday beaches. Every country has emergency plans to deal with such disasters. Here is an incident in 1999 when the French navy's ship L'Ailette equipped with vacuum pumps is approaching an oil slick threatening the French Atlantic

seaboard with an ecologial disaster. The oil spill was from the Maltese-registered tanker Erika, which broke in two in heavy seas in the Bay of Biscay.

Above: Desperate efforts are made to retrieve oil tankers before they shed all their cargo into the sea. Here in 2000 the rescue ship Smit Langkawi (front) tows the damaged oil tanker Natuna Sea through thick sludge in waters off Singapore. More than 7,000 tonnes of crude oil flowed into waters off Singapore and Indonesia from the Panama-registered

ship after the vessel ran aground near one of the world's busiest shipping lanes.

Above: When the oil from sunken tankers reaches the shore the clean-up is very messy, extremely difficult and often dangerous. Here volunteers clean up fuel oil on Muxia's A Pedrina beach in Galicia in northwest Spain on December 6th, 2002. Thick fuel oil was oozing from the sunken tanker Prestige off the coast and drifting onto the beaches.

Top: Physically carrying away the oil in buckets is frequently the best way of cleaning up since detergents often make pollution worse. Here two stained volunteers carry a bucket filled with fuel oil spilled from the same Prestige oil tanker. The ageing, single-hulled tanker foundered off the coast of Galicia in November 2002 with 77,000 tonnes of heavy fuel oil on board, causing Spain's worst ever ecological disaster, contaminating hundreds of miles of coast and putting thousands of fishermen out of work.

Above: Thousands of birds died like this one, unable to escape from the crude oil in the sea which leaked from the Prestige. Fish and wildlife were affected along hundreds of miles of coast.

"This is a rogue administration, out of step with the rest of the world and much of America, representing just a small number of powerful industrial interests."

Jonathon Porritt, chairman of the UK Sustainable Development Commission, December 9th, 2005.

the carbon club. He was recorded as having attended every one of the 20 meetings leading to the first Conference of the Parties of the Climate Change Convention in Berlin in 1995. He saw himself as a "preserver of the American way of life" and called his organisation the Climate Council. He operated out of the same building at the Global Climate Coalition but never revealed who his clients were, although his law firm represented, among 1,500 clients, Exxon, Shell, Texaco and Dupont. Pearlman was frequently seen in the company of delegates from the Gulf states, who wore the flowing robes of the desert, whether they were from Kuwait or Saudi Arabia. He was said to be more familiar than anyone alive with the 1,000 documents on climate change produced before the conference. He had argued about every line. He passed endless written notes to the men in robes, which observers claimed frequently turned into obstructive demands from the floor of the conference hall.

Unlike many lobbyists Pearlman was not concerned with the press. In his early years he would simply say "No comment" to any question about who he represented. After the Spiegel article he refused to be drawn on any subject, including what his aims or beliefs were. In later years he simply turned his back. He died in 2004, after single-handedly holding back progress on the science and politics of climate change for years. He did not live to see the Kyoto Protocol come into force. Most of his best work was done during the Clinton years, when there was some American impetus to doing something about climate change. He managed to stall progress long enough for an oil man to gain the presidency.

When George W Bush was elected the fossil fuel industry became deeply embedded in the White

House. It was not just that Bush himself was an oil man. Fossil fuel companies have part-funded his campaigns. Some of the White House staff were also former or actual lobbyists.

Harlan Watson, the US chief negotiator on climate, had been a friend of Pearlman's. It was Watson, as a Republican congressional aide in the early 1990s who urged the coal industry to hire Pearlman as a lobbyist. It was Exxon Mobil that suggested Watson be added to the administration's climate team. Perhaps just as significantly, Paula Dobriansky, Bush's under-secretary of state for global affairs since 2001, met Pearlman shortly after Bush pulled out of the Kyoto agreement. She said the purpose of the meeting was to "solicit [his] views as part of our dialogue with friends and allies". She also thanked Exxon for their involvement in formulating climate policy.

In June 2005, the extent of infiltration of the White House became clear when the New York Times revealed that Philip Cooney, a Bush aide and former oil lobbyist with the American Petroleum Institute, had doctored reports on climate change over a four-year period. Climate research from such internationally trusted and august bodies as the National Academy of Sciences, the National Ocean and Atmospheric Administration and NASA were altered and in some cases passages omitted, after they had been peer reviewed and adopted officially. Shortly after he was exposed he resigned and was immediately hired by Exxon Mobil.

Although the fossil fuel lobby has been remarkably successful in obstructing progress on dealing with climate change, the traffic has not all been one way. The Global Climate Coalition began to lose members rapidly as the 1990s drew on. Dupont, the US chemicals

Previous spread: It is not just old tankers which cause oil pollution, nor is it only wild-life and fishermen that suffer, the tourist trade can also be devastated. Here is an aerial view of an oil spill near Rio de Janeiro in January 2000. Brazil's state oil giant Petrobras said an underwater pipe ruptured, spewing about 500 tonnes of oil which had begun

washing up on beaches near Rio de Janeiro city. The Petrobras president, Henri Philippe Reichstul, called the rupture a "worrying acci-dent", which he estimated caused a moderate oil slick 3-5 km (2-3 miles) long.

Right: Exxon has become the target of protests across the world for its stance on global warming and its refusal to be repentant about it. The company claims that a boycott of its products launched world-wide has had no effect on sales. With a company that measures its daily profits in millions of dollars it is hard to tell. Lee Raymond, the

chairman and chief execu-tive, became a symbol for all that was wrong in the oil industry for environmental campaigners. He was dubbed the Darth Vader of global warming. There was an outcry in June 2006 when he retired after 12 years at the head of the world's largest oil company with a retirement package of $400 million (£214m).

During his leadership the company pumped an esti-mated six billion tonnes of carbon into the atmosphere, funded George W Bush's election campaign and many groups which deny global warming across the United States and elsewhere.

giant, saw new business in climate change technologies. The first oil company to withdraw was British Petroleum, which realised that the coalition's purpose to "cast doubt on the theory of global warming" might not fit in with its new solar business. The coalition began to collapse in March 2000 when Texaco defected but other main members, notably Exxon Mobil and General Motors, remain active in rubbishing climate change and the website remains full of misinformation.

Of course it would be wrong to ignore the fact that there have always been lobbyists from the other side. They have always worked extremely hard to persuade ministers from every country to take more action on climate change. They are not averse to infiltration either. Some of the European countries have included members of the Friends of the Earth and WWF as part of their official government delegations, perhaps helping to counterbalance the oil men in the giant US camp. There are literally dozens of non-governmental organisations, from one-man bands to large groups with 20 or more lobbyists. In fact the Greenpeace lobbying organisation at any talks is usually larger than 90% of developing country government delegations. Each group has its own point of view and some, like Greenpeace, would refuse to join government delegations to avoid compromising their position of being able to criticise anyone. Outside the conference halls too the insurance industry has always lobbied hard on its own account and backed fringe meetings and scientific presentations. Dire warnings on the escalating insurance claims because of weather-related incidents form the basis for reports about the parlous future of the industry's own member companies, as well as the planet, if global warming is allowed to continue unchecked. So the sceptics have not had it all

Top: Its sales badly dented by Japanese car makers producing small engined cars and hybrid vehicles, General Motors, the US giant, finally responded in 2006 by unveiling the new Saturn VUE SUV hybrid. The vehicle, which cost $25,000, still uses more petrol than the average European car.

Above left: Despite the continuing rise in oil prices and publicity about global warming there is still not much sign that US car makers have yet got the message. Here the 2005 Jeep Grand Cherokees sit on the assembly line during their production launch celebration at the Jefferson North assembly plant in Detroit.

Above right: In an attempt to cut costs Ford has gone over to robots to build their cars. Around 380 robots build the body of the new 2005 Mustang vehicle. Subsequently Ford has announced that it will compete in the smaller, greener car market with new models to try to recover some of its once worldwide dominance in the family market.

"Climate change is real. The science is compelling. And the longer we wait, the harder the problem will be to solve."

Senator John Kerry, opposing George W Bush for president in 2004.

their own way. The disadvantages and economic ruin the world faces because of climate change have been assessed and updated at meeting after meeting and made available to journalists and politicians.

So it is possible to argue that this is democracy at work, with all shades of opinion being represented. It is a valid argument. After all both the insurance industry and the fossil fuel lobby are trying to protect their own interests, and to a large extent the environmental organisations need to be seen to be campaigning to garner new members. However, when thousands of lives are already being lost because of climate change, which may soon turn into millions, should more be expected from lobbyists? The fossil fuel industry has been party to misinformation, obstruction and corruption. Warnings from the UK's Royal Society that this is still happening in the run-up to the release of the latest international scientific findings on climate change in 2007 show that the battle to rubbish the work of the Intergovernmental Panel on Climate Change still continues. It is hard to understand a motive, except greed, for all this time and money spent to discredit science and damage the political process. Some of the scientists involved with this lobby are no doubt sincere, there are after all still some legitimate uncertainties and doubts about speed and extent of climate change. But the main motive of those that sponsor, promote, and employ them, the corporations and the oil-rich countries, is to protect their interests. In other words it is about protecting profits at the expense of many lives.

Top: Many conflicts around the world are described as battles for freedom and democracy, or for the over-throw of dictators. This may be so but many of the most bitter and long drawn out have at their centre a power struggle over who controls valuable natural resources, and this is often oil. Perhaps it is a coincidence that as this Russian soldier lights a cigarette in front of a tank it is an oil pipeline burning in the distance. He is at a posi-tion on the outskirts of the Chechen capital Grozny. This was in 2000 when the then acting Russian presi-dent, Vladimir Putin, had just paid a surprise visit to Chechnya promising no let up in the fighting with Muslim rebels.

Above: Another of the world's oil hot spots, the Niger delta in Nigeria. Local people whose land has been ruined by pollution from oil exploitation have been involved in a long-running struggle with oil companies and the central government. Here in 2000 a fire contin-ues to burn along a ruptured pipeline in Oviri court in Adedje in the Niger delta after an explosion on the pipeline that claimed the lives of 250 people. Cash crops and acres of farmland were destroyed by the fire. It was the latest in a series of incidents in which people, who were believed to be illegally tapping into the pipelines, had been burned to death.

Following spread: Whatever the motives were for invad-ing Iraq, restored oil supplies were supposed to launch the newly occupied Iraq into a new era of peace and prosperity. The reality more than a year after the US-led invasion was a reduction in oil exports after the sabotage of the country's main inter-nal oil pipeline. The instability has led to further rises in the price of oil and the cost of the occupation has been a severe drain on the US economy. Policemen guard the burning pipeline near the city of Kerbala, around 110 km (70 miles) south from the Iraqi capital Baghdad on February 23rd, 2004, as the country gradu-ally descended into chaos.

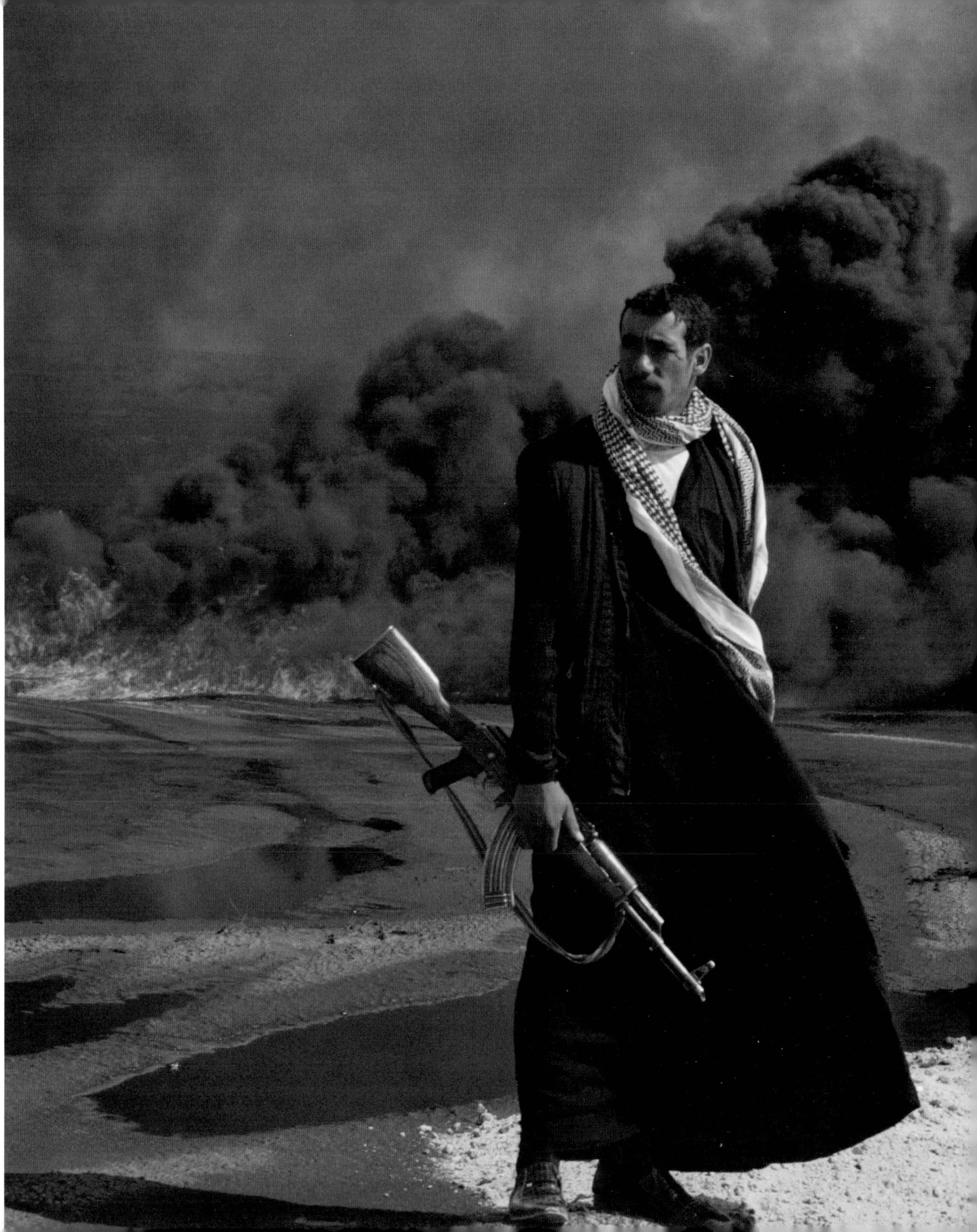

Even Termites are to Blame

Previous spread: Sunset over a rubbish dump in Guiyu, one of the most polluted districts in China. It is the final resting place of unwanted hi-tech waste from Europe and America. Electronic equipment, including televisions, computers and mobile phones, are the fastest growing toxic waste stream in the world. Although the export of such waste is banned under the Basel Convention, which all industrialised countries except the United States have ratified, much of it is dumped in China and other developing countries because it is claimed it is for recycling or reuse. Where electronic goods are dismantled by hand to obtain some of the precious metals inside the workers and communities involved are exposed to serious environmental and health problems from metals like mercury and lead.

Above: Living in the middle of one of the world's most troubled oil fields pollution incidents are frequent. This spurt of oil sprang from a supposedly sealed wellhead belonging to the Shell Oil facility at Kegbara Dere in Ogoniland, in Nigeria's volatile Niger delta in March 2006. Shell, which has been under pressure to improve conditions in the delta later admitted it had detected an oil spill at a dormant wellhead in the Ogoni area.

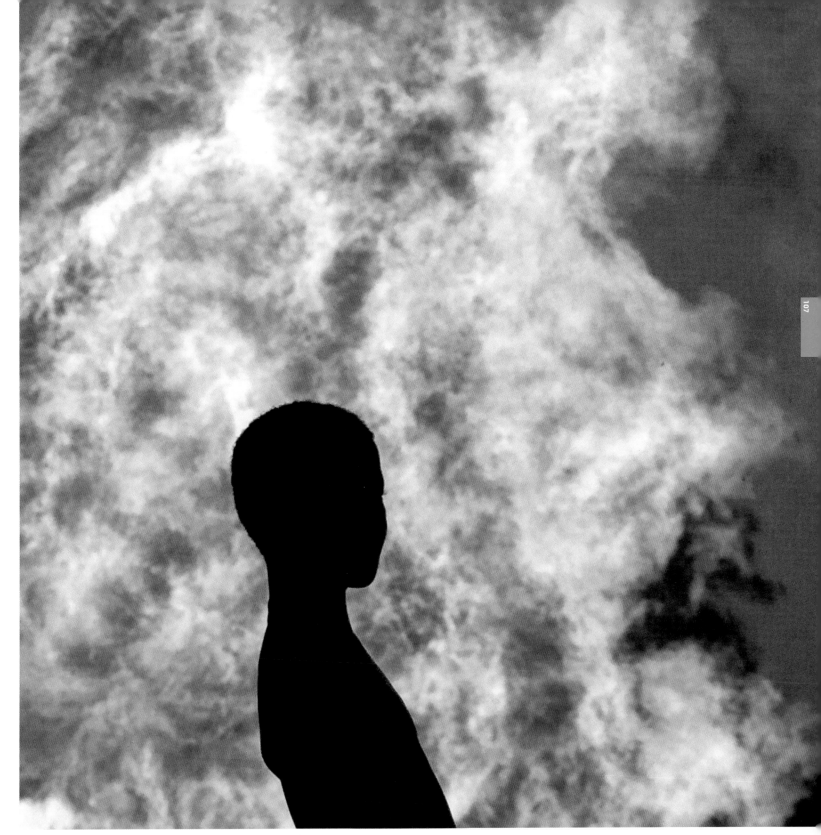

Above: Flaring gas is one of the most controversial activities in the middle of an oil field, since it is a relatively clean fossil fuel that could be captured and sold rather than burned. Here in 2004 a Nigerian child is silhouetted against a gas flare at the Shell Utorogu facility in Nigeria's southwest delta. This was taken on the same day that Nigerian troops dislodged hundreds of protestors. They had blocked access to Shell's facilities in the delta region and threatened to shut in 100,000 barrels of crude oil output.

There is no doubt that the temperature at the Earth's surface is rising and rapid changes are taking place around the planet as a result. Three massive reports running into thousands of pages, with experts from all over the world considering every aspect of the Earth's ecosystems, have confirmed this, along with one of the major causes of the changes – the burning of fossil fuels.

It has been the biggest scientific collaboration in human history, and has led to the agreement of thousands of scientists that the planet is in a parlous state. Yet somehow this process has still left room, for those who want to, to claim there is uncertainty and doubt.

As has been shown in the previous chapter the working of the Intergovernmental Panel on Climate Change, which has produced these three massive consensus reports over 15 years, has been obstructed and interfered with at every opportunity. Despite this, to the objective reader the reports make extremely gloomy reading. The contrarians, however, have achieved their purpose. Enough ifs, buts, maybes and uncertainties have been introduced into the text for every assertion in the reports to be subsequently questioned.

The headline facts in the 2001 report about the expected rise in temperatures between 1990 and 2100 are a perfect example. The report says the rise will be between 1.4°C and 5.8°C, a huge margin. It is low enough at one end to be "safe", that is below the dangerous threshold of 2°C discussed earlier, but well off tolerable limits at the other end of the estimate. Since this report was compiled new work suggests that these estimates are too low. Some scientists have calculated the upper limit should now be 11°C. But these summary figures of the possible extremes disguise the true magnitude of the problem, in two ways. First the average or what scientists think is the most likely average warming is somewhere between the two, about 3°C, in other words well into the dangerous zone where the "tipping point" is passed and man's ability to control climate change is probably lost. But the second misleading feature is that the increases do not reflect regional differences in temperature. In exactly the opposite way to

newspaper headlines that tend to alarm the reader before he has read detail in the text below, these temperature figures are averages across the globe and disguise what is going on in specific places. In general terms the temperatures on land are rising much faster than at sea, and so an average is misleading. This is particularly important, as will be seen in the next chapter, when looking at the Arctic and Antarctic, where the temperature rises have been particularly steep. These are the regions which are key to increasing sea level rise, the changes in the ocean currents, and release of massive stores of greenhouse gases locked up in the tundra. Western Siberia, for example, is heating up very fast. The temperature has risen 3°C in the past 40 years.

This and some of the other problems below are discussed in greater detail further on but it is important always to understand the inter-relationship between all these problems. Each one tends to make the next one worse, or to put it another way the changes happen more rapidly and with greater extremes. Ultimately they affect the ability of the human population to survive, especially in the current numbers, let alone the extra billions expected to be born between now and the middle of the century. Melting ice and tundra leads directly on to the next problem, and in the longer term the most intractable problem for the human race, sea level rise. Quite simply the land area on which we live and grow crops is shrinking and will go on shrinking for 1,000 years because of the warming we are already committed to. Not only are the glaciers and Greenland and other ice caps melting into the sea, the oceans are expanding as they warm. This and other knock-on effects of warm seas are becoming apparent. For example fish species, which have adapted to the ocean as it has been for thousands of years,

Above: Scavenging in rubbish is one of the last resorts of the very poor. Here Filipinos hike to the rubbish mountain called the Promised Land in Manila. It is dangerous work. Only days before this picture was taken in 2000 an avalanche of rubbish demolished over 200 shanties and 137 bodies were removed by rescue workers.

Top: This Indonesian scavenger has taken to a raft in a rubbish-laden canal in Jakarta on World Environment Day in June 2003. Many of Indonesia's impoverished live on the banks of the city's canals, which overflow during the rainy season.

Above: In South America the rubbish problem can also get out of hand. After flooding in 2003 caused 50,000 to flee their homes in Santa Fe, Argentina, these children were left to play among piles of rubbish heaped outside homes.

Above: An old man strains to pedal his tricycle into a sand storm which has turned the sky amber and greatly reduced visibility in Beijing, March 20th, 2002. Extensive deforestation and desertification in northern China have fuelled the dust storms. Nearly one million tonnes of Gobi desert sand blows into Beijing each year.

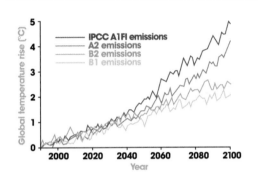

The four lines above show temperature rises over the next century. The rise depends on the quantity of greenhouse gases emitted in the first years of this century. Temperatures over land are expected to increase about twice as rapidly as temperatures over the ocean. The lowest line represents the increases forecast if emissions are reduced by 60% below 1990 levels. The top line is a continuation of current "business as usual". The lines in between represent some of the factors such as adoption of new technologies or oil running out which would effect emissions. Scientists say both of these scenarios are not nearly radical enough.

Source: IPCC/Hadley Centre for Climate Prediction and Research.

are very sensitive to temperature changes. The temperature of the water affects their ability to breed; just as important, the tiny creatures they feed on are even more sensitive to warmth and either disappear or move on. Oceans are also affected by increased acidity, as mentioned earlier. This is another potential disaster for the marine food chain and one of mankind's major sources of protein.

Although the droughts in southern Europe in the first years of this century, and the extreme heat wave of 2003, have alerted many in this otherwise green continent to the potential spread of deserts it is mostly in the developing world that loss of croplands has been causing alarm. One of the theoretical predictions of long-term climate change is that it will start to rain again in the Sahara but currently the world's largest desert is still expanding. It seems intent on crossing the Mediterranean to Spain, Italy and Greece. China is one of those countries where dust storms and desertification have been ringing alarm bells for years. A massive programme of tree planting is aiming to stem the tide. But although natural forces change the boundaries of deserts, and in some cases dry areas can be reclaimed by irrigation and careful planting, man is also responsible for creating deserts and enlarging them. Cutting down forests, overgrazing of dry lands, and poor irrigation practices which cause salts to build up in the soil have massively increased the area of desert in China. Climate change adds to these pressures and makes rehabilitating deserts more difficult, pushing even greater numbers of people to seek employment in the cities because they can no longer feed themselves on the land. Population pressure is one of the taboo subjects in the environment debate. But how do you continue to feed this ever increasing population in a warming world?

It was one of the great rows at the Earth Summit in 1992, when the Roman Catholic Church threatened to withdraw support if birth control was discussed.

Those issues are discussed elsewhere as we illustrate how the world is changing but it is also important to emphasise here the relationship of climate change to other environmental problems; the role of air pollution in its various forms, for example. The public, and even sometimes scientists, have been slow to understand the links. Acid rain is a good example of this interlinking. In the 1970s in Europe there was a long battle to understand and then at a political level take action to combat acid rain. All over northern Europe life in lakes was dying, even though they looked crystal clear, and pine trees were losing their needles. Whole forests were denuded. The culprit was low quality coal being burned in power stations. The coal produced noxious fumes in the form of sulphur dioxide, which in dry weather was deposited directly onto the ground, but often mixed with the clouds and produced a weak solution of sulphuric acid. Smoke from British power stations drifted hundreds of miles to the northeast and fell as acid rain in southern Scandinavia. The pattern was being repeated all over Europe and the world. The damage was not just to fish and trees. Acid soils changed the type of plants able to survive in certain areas and damaged buildings; for example gargoyles on cathedrals carved of limestone literally dissolved. The problem of acid rain is solvable and rich communities like those in Europe and North America set about reducing sulphur emissions by filtering smoke before it was released. This had a surprising side effect. Unlike carbon dioxide, which effectively remains in the atmosphere for around 100 years, sulphur dioxide lasts only a few days or at most weeks. As soon as the smoke stacks began to be cleaned up, the air cleared, and the temperature began to rise. It was studying the Mount Pinatubo eruption of 1991 and the 20 million tonnes of sulphur dioxide it put into the atmosphere that confirmed to scientists studying global warming that pollution had been holding down global temperatures by reflecting back sunlight. Some scientists now believe that the huge level of pollution between the 1940s and the 1970s held world temperatures down when they would otherwise have been rising in response to increased carbon dioxide levels. If this is true then it adds to the scariness of current climate change. Some of the NASA pictures from space show massive pollution either side of the Himalayas. This is from the surging fossil fuel use in India and China, which has been accelerating over the last 30 years. This too must be greatly limiting the rise in temperatures which is already apparent. These two countries are both attempting to tackle the life-threatening pollution in their cities. If they succeed they will send the temperatures soaring over Asia. Another complex interaction between two apparently unrelated environmental problems is the role of ozone depletion. The well-documented "hole" in the ozone layer caused by the release of man-made chemicals into the atmosphere is a problem because it allows larger quantities of harmful ultraviolet light to reach the earth. It also alters the balance in the amount of heat retained by the atmosphere. While greenhouse gases heat the lower layer, called the troposphere, ozone depletion caused the stratosphere to cool. While ground stations all measured the earth heating up, satellites, which give more of a global average, showed virtually no warming. It provided a field day for the contrarians, throwing doubts on all of the

Top and above: Everyone in China has been called in to take part in the country's vast tree-planting programme. Disastrous floods in recent years and evidence of rapidly spreading deserts have both been blamed partly on the removal of trees in the mountains. Pensioners, soldiers and children plant trees. In the top picture sol-diers dig the holes while schoolchildren carry the saplings to be planted. This and the picture of a child resting between planting sessions were taken 70 km (45 miles) from Beijing where belts of trees are being planted as a protection for the capital.

Right: Where new forests have apparently been successfully replanted the Chinese are taking no chances. The plane is on an air seeding mission in Lushi, central China's Henan province. In 2005 the local government conducted air seeding over 6,500 hectares (16,000 acres) and claims to have reduced the area of desert under their control.

However, in China as a whole the total land area affected by deserts is said to increase by 2,000 square kilometres (800 square miles) a year.

science. Some scientists then worked out that the satellites had been averaging out the warming of one layer with the cooling of another. Discovering precisely what is happening remains elusive and this issue remains a hotly contested area.

So it seems that although air pollution, acid rain and the ozone depletion are all serious problems, which must be solved because each one has detrimental effects on human health and the environment, their cure will also have a bearing on how fast the earth heats up. In some heavily polluted areas in Asia the effect could be large. Although the net result of civilisation's attempts to solve these various environmental problems on climate change is still not quantified, there have been some more recent attempts to discover how much the climate will warm. As discussed earlier there is an emerging consensus that anything above 2°C risks disaster. It is also a tall order for many long-lived species like trees to adapt to a rapid rise in temperature. One degree celsius rise is equivalent to moving 150 miles south in Europe and a similar rise would move the tree-line 150 metres (500 ft) up a mountain. In early 2005 a team from Oxford University used the power of 90,000 linked personal computers to run a series of trials to see what effect a doubling of carbon dioxide would have in the atmosphere — that is around 550 parts per million. As we have already discussed, the temperature this level of carbon dioxide produces has an important bearing on the political decisions that need be taken to avoid dangerous climate change. This 550 ppm level of carbon dioxide in air is also one that the IPCC scientists have been working on to make their predictions, because on current trends, that will be the level by 2050. The resultant predicted increases ranged between 1.9°C and 11.2°C, depending on how

different processes in the atmosphere were represented in the models. This is a wider range than the previous IPCC results, but a revision upwards. Even the lowest figure is perilously close to the 2°C danger threshold. The effect of 11.2°C would be very dramatic.

An important aside here is that throughout, the amount of carbon dioxide in the air has been central to the scientific and political argument. Methane has also been mentioned as a key greenhouse gas. There are other greenhouse gases including chlorofluorocarbons, or CFCs, which are being phased out anyway because they are the major cause of the destruction of the ozone layer. As has been mentioned earlier, there were six greenhouse gases, or groups of gases, mentioned in the Kyoto Protocol, the reduction of all of them counting towards any country's target. Apart from carbon dioxide and methane these are nitrous oxide, otherwise known as laughing gas, and two groups of chemicals called hydrofluorocarbons and perfluorocarbons, which are produced in chemical manufacture, and sulphur hexafluoride. All of them have a far greater heat-trapping effect in the atmosphere, per kilogramme emitted, than carbon dioxide but are in such tiny quantities that properly controlled they will not be a problem. Nitrous oxide is released with the production of nylon and is produced naturally in the soil but excessive fertiliser use is a major human cause. Extensive efforts are being made by industry to control the other industrial gases and find non-damaging substitutes.

Methane is very significant because it has 23 times the global warming potential of carbon dioxide per kilogramme emitted and is produced in large quantities both naturally and because of human activities. Natural sources include rotting vegetation in wetland. Even

Left and above: Acid rain has damaged forests and stone monuments across large areas of Europe and is an increasing problem where coal is the main energy provider in the developing world. The acid comes from sulphur in coal and oxides of nitrogen producing weak solutions of acid when smoke mixes with the clouds. An extreme case is the dead spruce trees in the Ore mountains, the Czech Republic, but more typical is the dissolving of this statue in Krakow, Poland. Many old buildings were built with limestone, cemented together with limestone cement. Acid rain dissolves both stone and mortar, resulting in irreparable damage. Statues made of marble, a form of lime- stone, are also susceptible to acid rain. The Acropolis in Athens was damaged more in 25 years than in all 2,400 years before. It costs Europe $9 billion per year to replace corroded stone because of this man-made problem.

"There may be technology to deal with emissions from the industrial and energy sectors but we have not yet found ways to stop cows and sheep from doing what comes naturally."

David Parker, New Zealand's minister responsible for climate change issues, told a climate change conference in New Zealand in 2006, referring to his country's millions of flatulent livestock.

Above and right: Rice is a thirsty, labour-intensive crop and a staple food across large areas of the world. Climate change is threatening the ability of large areas in China to grow enough rice to feed the population and everywhere peasants are leaving the land to move to the cities. China expects to use its new economic power to import grain from its neighbours but for how long? Vietnam is still a net rice exporter, with the grain grown in the traditional manner. Above, a farmer, with ancestors' graves behind and new rice seedlings in front, turns his buffalo as he ploughs to prepare a new paddy field on the outskirts of Hanoi. In Bangladesh boys (right) also follow tradition, washing paddy seedlings at Singair near the capital Dhaka, but the country is also vulnerable to extreme weather which threatens the food supply. In 2004 2.77 million acres (1.1 million hectares) of crops were submerged by floodwater.

termites burp 20 million tonnes of methane a year, more than is released from the oceans. Man-made sources, in particular coal mining, natural gas and oil exploitation, release far more — about 100 million tonnes. Troublesome quantities also come from rice paddies and the digestive systems of cows and sheep. One large source in Britain is landfill sites; a new industry has developed to capture and burn the gas as it is released to produce electricity. Power from landfill gas is one of the cheapest forms of electricity and has been one of the key examples of the benefits of tackling climate change. Although burning the gas produces carbon dioxide, this has far less global warming potential than the original methane. Although methane remains an important greenhouse gas, direct man-made emissions of the gas are being successfully reduced in many countries. The most important problem remains carbon dioxide simply because it is the gas we are pumping out in the largest quantities and seem politically unable to tackle. Potentially, however, it is the easiest problem to deal with. All we have to do is stop using fossil fuels. Man's rapid development over the last 200 years, which brought the industrial revolution and a population explosion to the planet, has also brought society to this crisis. Science and politics have intertwined and clashed. Alongside this there has been a long argument, which still runs through most religions, about man's dominion over nature versus the responsibility for stewardship of God's creation. It has important political implications both in the Muslim world and in the influence of the religious right on the Republican party. Both the Orthodox Christians and the Roman Catholic Church have started taking the environment seriously, and protecting the environment has become part of both churches' teaching. But as far as most of mankind is concerned, the forests

Above and right: Water is the biggest potential point of conflict between the three countries that share the great rivers that flow south from the Himalayas. With the water supply threatened, partly because many of the glaciers that feed the rivers are disappearing, tensions are bound to grow. There are agreements between the three — India, Pakistan and Bangladesh — about sharing the water from rivers but any new project is controversial. For example on the Chenab river (above), the Baglihar hydroelectric project led Pakistan to accuse India of violating the Indus Water Treaty by restricting water flow to generate electricity. The scheme is in the troubled northern state of Jammu and Kashmir. The World Bank is mediating. The picture (right) from space shows how interdependent the countries of the region are. The semi-arid Tibetan plateau (upper left) meets up with the Himalayas to the south. From the heights of the Himalayas, snow-covered on their northern flanks, and lush with vegetation to the south, numerous rivers, brown with churned up sediment, flow into the valley of the Brahmaputra river in Assam, India. The Brahmaputra turns southward at the border of Bangladesh and is soon joined by the Ganges river. This river splits into numerous channels as it runs out toward the Bay of Bengal, giving the region the name "Mouths of the Ganges".

and fields and the animal and fish stocks have always been treated as natural resources which are infinite, even if there is plain evidence to the contrary. If the clean water runs out, pipe some from over the hill. If you chop down a forest, import timber from elsewhere. No matter that north Africa was once forested until the Romans cut down the trees to grow grain or that most civilisations that have disintegrated have done so because of misuse or lack of natural resources. The magnificent but lonely statues of Easter Island are a classic example of a civilisation which became impoverished because it cut down all the trees, could no longer build canoes to hunt food, and simply ran out of the ability to sustain itself.

All that was a long time ago, but it is obvious that so far few lessons have been learned. But with the aid of constant monitoring from space mankind can clearly see on a daily basis how his activities are affecting every part of the globe, often disastrously. Forest fires, droughts, even the progress of new logging roads through the Amazon can be seen from satellites. It is also clear that none of these things are happening in isolation. The logs cut in tropical forests, mostly illegally, end up as furniture or expensive building materials in the rich world. The pollution emitted in India or China is to power factories, which export cheap goods to Europe and America. The pollution in those far-off places and in Europe and America caused by our profligate lifestyles chokes both our and their citizens. At the same time it keeps the temperature lower than it would otherwise be. In the next chapters we look in more detail at some of the consequences of what man has been doing.

Above: The Ganges is a sacred river used for Hindu ceremonies. It is also a great life-giving supply of water and therefore food to millions of people, all of which could be in jeopardy. It is fed by snowfall in the Himalayas and the rain from the monsoons, both of which are changing with the warming climate.

Right: Thousands of Hindu worshippers bathe in the holy Ganges in Allahabad, February 8th, 2001, during one of the most auspicious days of the Maha Kumbh Mela festival called Maghi Poornima. Tens of millions of Hindus gather over a six-week period to bathe in the Holy waters of the Ganges at the festival which is held only once every 12 years.

We Need Ice

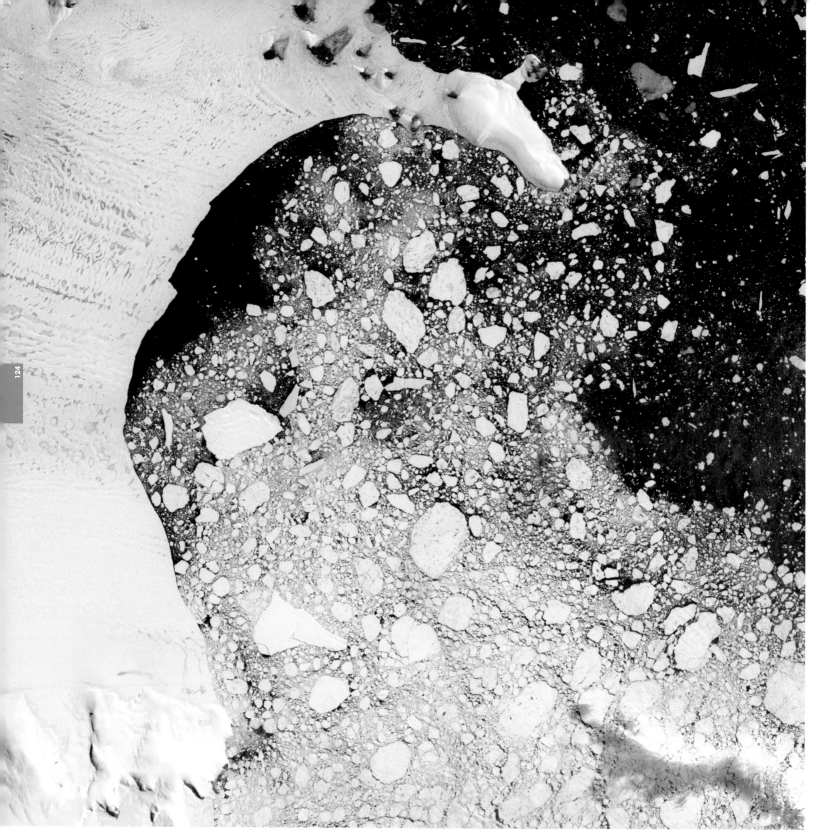

Previous spread: Northern Canada and Greenland seen from a satellite, in summer 1999, are still predominantly white as ice and snow continue to cover most of the land and sea. This image will be used in the future to show just how much global warming has affected the ice cover on the roof of the world.

Above: The Larsen ice shelf in Antarctica as seen from space. Warmer temperatures over a few months can cause huge areas of apparently stable ice to collapse. Scientists can see pools of melt water on the surface, which fills crevasses and increases the pressure on the shelf, extending the cracks, and causing it to disintegrate.

Above: This is not the normal quality of image hoped for in a book like this but this picture taken in 1997 of hikers on top of the Rhône glacier in the Swiss Alps is unrepeatable. It has retreated so much that they would now be walking on air. The glacier is the source of the River Rhône, which feeds Lake Geneva before running across France to the Mediterranean.

Ice is crucial to the way Earth's climate works. It is also vital to the survival of many wild creatures and provides the regular water supply for hundreds of millions of people.

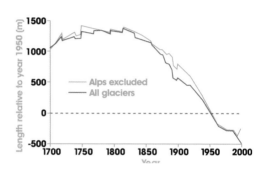

Glacier retreat is probably the most obvious and easily understood proof of climate change. The curve above represents measurements of glaciers on which there is reliable information available; it shows that in two centuries these ancient rivers of ice have retreated an average of nearly 2,000 metres. For the period between 1900 and 1980, 142 of the 144 glaciers for which adequate data was available decreased in length. The exceptions were in areas where climate change has caused a considerable increase in snowfall, replenishing the glaciers faster than they melted. Since 1980, in rapidly increasing temperatures, glacier melting has accelerated across the world.

Source: H Oerlemans, Utrecht University

But as temperatures rise from Kilimanjaro near the equator in central Africa to Alaska, glaciers are disappearing. In the highest mountains in the Himalayas, around the Greenland ice cap, even at the North Pole and in Antarctica, the accumulated ice of millennia is melting. This is not the only concern. On high mountains and across vast tracts of land in northern forests and treeless tundra the land is frozen underground, and has been since the last ice age. More than a quarter of all the world's land area is what is called permafrost. Many of these areas are melting too.

Perhaps the first thought that springs to mind about melting ice is its effect on sea level rise. This is already causing alarm in many parts of the world — an issue dealt with in the next chapter. The other effects of melting ice are not so immediately obvious but are equally important and for many, far more devastating. The world's food supply is heavily dependent on the winter build-up of snow on high mountains which is compressed to form ice caps and glaciers. Each spring some of the snow melts in a rush and then during the dry summers the glaciers provide a smaller but reliable and constant top-up to feed the rivers below.

To understand the increasing impacts of melting ice it is important to understand the role of ice and snow on the climate. It is calculated that about one third of the sun's rays are reflected back into space after striking the earth. The amount of light reflected is not uniform. White ice and snow reflect about 90% of the light and most of the heat. Ocean water on the other hand absorbs most of the light and heat. As the world warms and the area of ice and snow declines, the areas around the edge, both on land and at sea, absorb quantities of energy from the sun which was previously sent

straight back into space. This warming, which then causes more warming, is called a positive feedback. This partly accounts for some of the dramatic changes seen in the Arctic in the last 20 years since satellites have been photographing the earth. In 2005 experts at the US National Snow and Data Centre in Colorado said that in September that year, the maximum time of ice melt, the extent of Arctic sea ice had dropped 20% below the long term average. An extra 500,000 square miles, or an area twice the size of Texas, had turned from reflective surface ice to water. This was the fourth successive year that melting had been greater than the average.

Walt Maier, one of the scientists, said: "Having four years in a row with such low ice extents has never been seen before in the satellite record." The melting ice almost completely cleared the notorious north-west passage of ice. This area to the north of Canada is where many heroic expeditions were lost in past centuries trying to find a new shorter sea passage from Europe to Asia. Soon it will be possible to sail unhindered along this route. Already the north-east passage across the top of Siberia is open in summer. The satellite measurements also show that in winter the ice never re-forms over such a wide area as it once reached. The spring melt also begins earlier; in 2005 it was 17 days earlier than the 20-year average.

Shortly before the announcement about the loss of ice on the seas a joint research project by Russian and British scientists revealed a remarkable melting of the permafrost in Siberia. It is the world's largest peat bog. An area the size of France and Germany combined was developing lakes and turning to mud for the first time since it formed 11,000 years ago. Western Siberia is heating up as fast as anywhere else in the world, recording a 3°C rise in 40 years. The

"It seems to me that the natural world is the greatest source of excitement; the greatest source of visual beauty; the greatest source of intellectual interest. It is the greatest source of so much in life that makes life worth living."

Sir David Attenborough, in interview with the BBC, 2006.

Above: Shortage of snow in the Alps has caused a crisis at some winter sports resorts. Increasingly snow-making machines are used in many places but they are very expensive to run and add to climate change because they are powered by fossil fuels. To try to get round the problem, the authorities here at Mayrhofen in Austria give skiers and snow sport enthusiasts a pass, which allows them to move to another resort.

biggest fear of scientists is that some of the 70 billion tonnes of methane trapped in the frozen soil will be released by this warming. Since methane is 23 times as potent a global warming gas as carbon dioxide this is another potential large scale positive feedback. David Viner, from the University of East Anglia, said at the time "This is a big deal because you cannot put the permafrost back once it's gone. It will ramp up temperatures even more than our emissions are doing."

The parts of the planet where these changes are taking place are generally the least inhabited. There are European, Russian and Canadian tribes with settlements well inside the permafrost zone who have been forced to move because their homes have begun to tilt and collapse as the soil has melted. Oil and gas pipelines have also suffered leaks as their permafrost foundations, once thought to be as solid as rock, have sagged. Part of the Trans-Siberian railway has also had to be rebuilt. The melting has also caused what are called drunken forests, where the trees lean over at crazy angles as the ground has sunk unevenly.

The tops of mountains and the edges of the Arctic circle where the ice and permafrost occurs are frequently uncomfortable places to live, at least for humans. However, as on every part of the planet, there are specialist animal and plant species that have adapted to even the most inhospitable climate.

Perhaps the best known example of a species that has evolved to make the most of the ice in the Arctic is the polar bear. Apart from the obvious camouflage of the whiteness of its fur and its ability to hibernate underneath the snow, the polar bear depends for its survival on the ice. Its main winter food is seals and the

Left: Animals have evolved and adapted to occupy habitats that are extremely inhospitable. The polar bear is the top Arctic predator and its main diet is seals, which it catches on the sea ice. The disappearance of the ice, particularly its early melting in spring, is making it difficult for these bears to feed their young and is threatening their survival.

Top: The walrus does not get as much publicity as the polar bear but it too is a specialist Arctic dweller and survivor in extreme conditions. It too faces life-threatening changes to its hunting and breeding grounds. In the background is the Greenpeace ship Arctic Sunrise in Chuckchi Sea investigating climate change in the region.

Above: The great migration of the caribou in their thousands across the Canadian tundra to their calving grounds is one of the wonders of the animal kingdom. The traditional migration routes are threatened both by oil exploration and by the changing climate.

hunting ground is the ice where the seals breed and fish through holes in the ice. The bears can swim long distances but not well enough to catch seals in the open water. If the ice fails to form then the bears cannot hunt and may starve to death.

Scientists believe that polar bears will disappear across most of their previous range in the northern regions as a result of the shrinking Arctic ice cap and may become extinct in the wild in the next 50 years. At the opposite end of the world, the Antarctic, the disappearance of the ice is already having a dramatic effect on wildlife. A shrimp-like creature called krill is central to the diet of fish, penguins, seals and whales. In the summer there can be swarms of krill so large that the sea appears pink. Families of whales can eat millions of them without appearing to make a difference to the vast swarms. But research shows that krill numbers have dropped dramatically since the 1970s and scientists believe that the loss of sea ice is the explanation. Krill feed on the algae found under the surface of the sea ice in winter. The ice acts as a shelter for the krill, which have a chance to thrive before it melts and exposes them to predators. Scientists from nine countries working in the Antarctic pooled their data on the species and concluded in 2005 that numbers had declined 80% in 30 years. This is a dramatic loss in the main food supply of species like penguins. Scientists believe this accounts for the depletion in penguin numbers because they rely heavily on a plentiful supply of krill to feed their young in the short Antarctic summer.

But krill and their predators are not the only creatures affected. The Southern Ocean is about 1°C warmer than in the 1960s and a lot of bottom-dwelling creatures like molluscs, limpets and scallops are struggling to adapt. Strange

Above: Killer whales or orcas, travelling in family groups, have an amazing range across the world's oceans and eat almost anything they can catch. Food would include seals, seen here (top) on a melting glacier in Prince William Sound, Alaska.

Right: In the Antarctic these Adelie penguins stand framed on a massive iceberg. The ice is also where young penguins can grow large enough to have a reasonable chance of survival before taking to the dangerous waters. As the ice melts this haven will disappear.

"This is the biggest challenge our civilisation has ever had consciously to face. If this goes on we will lose ice cover on our planet. The process will cause such rapid transformation we will have enormous trouble adapting."

Sir David King, the UK government's chief scientific advisor, London, March 2006.

though it sounds, limpets cannot turn over in warmer water and scallops lose the ability to swim. The problem for these creatures is that they cannot move further south to find better conditions; the giant land mass of the Antarctic stands in their way.

It is elsewhere on the planet that the melting ice directly affects humanity. Kilimanjaro gets its name from the Swahili for shining mountain, because of its snow cap; it shines no more. The highest mountain on the African continent still attracts the tourists because of the fantastic wildlife around its foothills and the beautiful forests on its flanks. But the glaciers and the snow which once crowned it are almost gone. Probably within 10 years the crater of this fantastic volcano will have lost its ice for the first time in 10,000 years.

All over South and North America, Europe and Asia glaciers are melting at an ever faster rate. In some places like the Alps, where skiing is a major industry, it has an important economic impact. In the Italian Alps during the heat wave of 2003, 10% of the total mass of glaciers melted in a single summer, and experts fear they could disappear completely in 30 years.

In Europe, Africa, Asia and the Americas glaciers are vital for another reason — water supply. To people sometimes hundreds of miles downstream it is not obvious that most major rivers in the world are kept flowing by underlying glaciers once the spring snow melt has finished. These glaciers are vital in arid regions for releasing water through long dry summers.

Over the last half century as glaciers have generally melted at an ever faster rate the water supply in the valleys below, sometimes far away,

Left: Elephants and zebras in front of Africa's tallest mountain, Kilimanjaro. The name Kilima Njaro means "shining mountain" in Swahili, but its once permanent ice cap, which is such an attraction for tourists, is melting fast because of climate change. Scientists say that the glaciers, which have been there at least 11,000 years since the end of the last ice age, will be gone within 15 years and then the occasional snow will melt not long after it falls. The rain falling on the lower slopes of the mountains is important to feed the springs and rivers that provide water for the big game and people who live on the plains. The loss of ice will bring unknown changes to the hydrology of the area.

134

1998

1998

2003

2003

Above: These pictures document the loss of glaciers on the top of Mount Kilimanjaro. This mountain, which dominates its surroundings, is remarkable because of the climatic changes in such short distances from the dry tropical plains, through rainforests to cold grasslands and finally to the icefields at the 5,895 metre (19,340ft) summit. In the last 40 years 55% of the ice in the glaciers has disappeared and the process is speeding up dramatically, as these pictures taken in 1998 and 2003 show. Since the mountain is only 200 miles from the equator there is no summer or winter but the pictures are taken at the same season. In the past five years vast amounts of ice have disappeared. The mountaineer who took the photographs, Alastair Dobson, said of his experience: "I didn't recognise the top of the mountain in 2003 — gone were the walls of ice my daughter and I had rested against in 1998, gone was the corridor between the cliffs of a glacier — all I could see was volcanic ash, boulders and rocky outcrops. The trail leading to Uhuru looked like a beach — no snow, no ice, just a broad expanse of brown gravel."

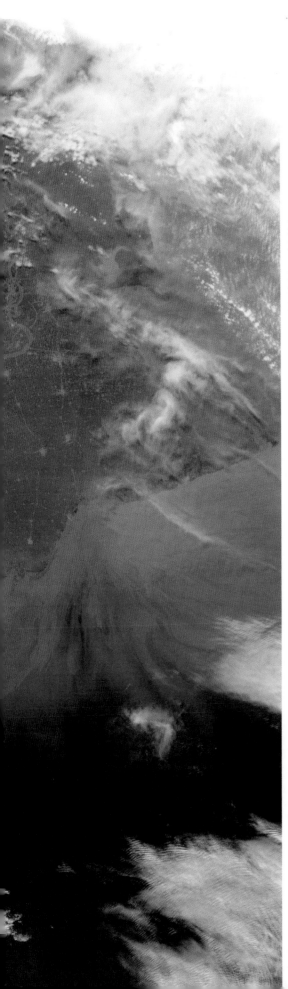

has often seemed ever more plentiful. But this bonus cannot last; the ice is disappearing, and there are hidden dangers far up the valleys.

For example the great rivers of Asia, which provide water for the most populous regions on earth, are fed by the glaciers of the Himalayas. The Ganges, Indus, Brahmaputra, Salween, Mekong, Yangtze and Huange He all owe their summer flow to the melt water of glaciers. They provide water for drinking, irrigation and hydroelectric schemes. As glaciers have pushed down the mountains through the centuries they have ground out rocks and earth into giant mounds along the their front edges and along their sides. As the glaciers recede these mounds of rocks and debris, called moraines, act as dams and create freshwater lakes where there was once ice. These can be incredibly dangerous because the dam walls are often partly frozen at the base and as they melt the whole structure collapses. In the steep Himalayas this can lead to catastrophe, as a wall of water, mud and rocks hurtles down the mountainside. In 1985 a glacial lake burst in Khumbu, Nepal, killing at least 20 people. It also washed away a hydropower station, a trekking trail and numerous bridges.

Latest surveys of 3,300 glaciers in the Nepalese Himalayas show 2,300 of them have glacial lakes in the valleys where they have receded. These lakes are continuing to grow because of rising temperatures. Jennifer Morgan, director of the WWF's global climate change programme, says: "Himalayan glaciers are among the fastest retreating glaciers globally due to the effects of global warming. This will first increase the volume of water in rivers, causing widespread flooding dangers. In a few decades this situation will change. Water levels in rivers will decline, eventually resulting in water shortages for hundreds of millions of people who rely on glacier-dependent rivers. This threatens massive economic and environmental problems for people in western China, Nepal and northern India."

When I visited Georgia in spring 2006, the villages on the southern side of the Caucasus had been placed on alert because of the danger of flooding and mudslides. Although part of the problem was deforestation because impoverished people had cut down the trees to try to keep warm, climate change was also said to be to blame. The homes of 400,000 families in 3,000 villages were at risk, according to Professor Emi Tsereteli, of the State University of Georgia, who heads the country's ministry of environment's centre for the "prognosis and diagnosis" of disasters. In the previous 35 years the glaciers of Georgia had retreated 27%. They had an important role in the summer flow of two of the country's major rivers, Enguri and Tergi. An example of the changes was that the Alpine forests were now growing 200 metres higher in the Caucasus.

Professor Tsereteli said that as a result of the changes "mud flows, landslides, flooding and other gravitational processes are costing Georgia at least $150 million [£80 million] a year in damage, and it could easily reach as much as one billion". He said this was not a new problem for Georgia. "Over the last 30 years 50,000 families have been relocated because of environmental disasters but every year Georgia is facing a worsening situation." One of the problems he most fears is mud slides caused by melting glaciers, which cut into the hillside. They had the biggest potential for a catastrophic accident like the one on the Caucasus in southern Russia that was caused by the collapse of the Kolka glacier in September 2002. An avalanche of ice, snow and rocks hurtled eight miles and

Left: China's battle to control the flow of its rivers, in this case the Yangtze, begins in the Himalayas and ends here in the Yellow Sea where the sediment-rich plume can be seen from space. The snow melt and glaciers account for the spring flood and much of the flow of the river in summer. As the glaciers recede further and further each year it appears far downstream that the summer flow is plentiful but when the mountains warm sufficiently and the ice finally disappears the water supply to millions will be put in jeopardy.

Left: An early example of what can be done to improve the quality of life in the developing world without damaging the climate. The Himalayan village of Namche Bazar in Nepal has had electricity since 1983 when the Austrian government paid for a hydroelectric project so that light bulbs could replace kerosene lamps. The village is surrounded by towering peaks, which are the watershed for the rivers that bring life to China on one side and India, Pakistan and Bangladesh on the other.

Top: Scores of Cambodian fishing boats gather at dawn on the Mekong river for the high-point of the fishing season on January 4th, 2004. Huge fish migrations every January, caused by the annual flood cycle of the Mekong, provide a vital source of protein to Cambodia's 13 million people, although conservationists fear that upstream damming threatens the long term future of the delicate ecosystem. The health of both rivers and their flow are directly affected by the climate changes in the mountains.

Above: Participants with life buoys and balloons swim across the Yangtze river at the opening ceremony of the 2005 China Wuhan International Yangtze River Crossing Challenge in central China's Hubei province, on June 18th, 2005. More than 1,100 enthusiasts from all over China took part in the 2,000 metre floating contest at the ceremony.

"Besides retreating glaciers, insect infestations and more intense forest fires, Alaska is experiencing melting permafrost, flooded villages, warming oceans, coastal erosion, shifts in bird and wildlife populations, and shorter seasons for ice roads. And there is more to come, as Alaska is heating up at twice the rate as the rest of the world."

Kate Troll, an environmentalist writing in the Anchorage Daily News on the tourist boom to see the state's glaciers before they disappear, August 2005.

swallowed the village of Nizhniy Karmadon and other settlements killing at least 125 people. This could just as easily happen on the Georgian side of the Caucasus, he said. He also thought the highway that runs north through the mountains linking Georgia with Russia could be cut by landslides, as could a gas pipeline along the same route, which is vital to the region's energy supply. The pipeline had been threatened by two landslides in 2005.

At the other side of the world, in Peru, the story is the same. Using the vast Andean ice caps and glaciers, 70% of Peru's power comes from hydroelectric dams catching runoff, but officials fear much of it could be gone within a decade. At the same time new mountainside lakes are bulging from the melt, threatening to break their banks and devastate the towns below.

But this is a world-wide phenomenon and there is no way that even the richest and most powerful can shield themselves from the effects of climate change on ice. California, the sunshine state, probably has the most sophisticated large-scale water control systems on the planet. It also has a carefully studied climate. How much it snows in the Sierra mountains each winter means a lot more to the 36 million Californians than how good the skiing will be. The spring and summer meltwater is carefully collected and shared out. It is vital to industry, agriculture, domestic users and the wildlife. California's department of water resources has already predicted that demand would exceed supply by 2020 — and the shortfall would be the equivalent of a year's supply to 11 million people. That was before a report by the National Academy of Sciences in August 2004, which said that climate change would cause a serious decline of the Sierra snow pack. The report said that depending on how much world-wide emissions of carbon

dioxide were cut, warming would reduce the winter snow by between 30% and 90% by the end of this century. The runoff from streams would drop by at least half.

The 19 researchers from leading US universities studied the snow pack because they said the volume of water produced by the melt affects 85% of the drinking and irrigation water available for Californians. In addition snow-fed hydroelectric plants currently produce about a quarter of California's power. This knowledge about how much carbon dioxide emissions will affect local climate has caused Arnold Schwarzenegger, California's governor, to adopt an 80% greenhouse gas reduction target, the most ambitious in America. But in most areas of the world this political action is missing. This is probably because the direct connection between global warming and the water supply is not made, and so the sometimes imminent threat to the future wellbeing of local people is still not appreciated.

Above: This is what every Californian wants to see each winter, and it is not just for the skiing resorts outside Los Angeles. This picture was taken in December 2002 when 12ft (3.5 metres) of snow fell in the Sierra Nevada mountains. The annual snowfall is the life-line for the sunshine state's vast agricultural industry and the spring runoff is carefully collected in dams to provide the summer water supply for farmers and the people of a state that would otherwise be parched.

142

c.1920

c.1920

2002

2002

Above: A series of before and after pictures documenting the retreat of glaciers in Svalbard, Norway. This is well inside the Arctic Circle and the most northerly point washed by the warmer waters of the North Atlantic Drift. The water begins its journey in the Gulf of Mexico and brings heat to western Europe as far north as Svalbard where the sea temperature in 2006 was a full 2°C warmer than normal. The pictures at the bottom were all taken in 2002 when a Greenpeace expedition went to see if they could put volunteers in the landscape in the same place as the Norwegians were in the originals above them. All the pictures are of different views of the giant Blomstrandbreen glacier in Kongsfjorden. The earliest on the far right was taken in 1918, with all the rest taken in the 1920s. Large areas of black rock covered since the last ice age 11,000 years ago are now absorbing heat from the sun, adding to global warming and accelerating the ice melt.

Rising Tides

Previous spread: The hurricane season begins in June and now lasts almost half a year in the Caribbean and along Florida's coast, with 2005 being the worst storm season ever recorded. Here Key West resident Gregorio Nodal walks in the flooded North Roosevelt coast boulevard after hurricane Wilma hit Florida's southern west coast on October 24th, 2005, two months after Katrina devastated New Orleans. Hurricane Wilma crashed ashore in southwest Florida, again bringing flooding before roaring across the peninsula, pounding Miami, Fort Lauderdale and West Palm Beach. It had previously hit Mexico's Yucatán peninsula and killed 17 people across the Caribbean.

Left and above: The locals call them King Tides and their children see them as fun as they play in the incoming sea as it sweeps across their islands. But many fear it is the beginning of the end for Buota village, on Tarawa island, one of the many vulnerable settlements on the scattered atolls of the nation of Kiribati, in the Pacific Ocean. These low lying coral islands, which currently are only flooded rarely, are expected to be completely inundated as sea level rise continues. These pictures taken in February 2005 show that preserving fresh water supplies and growing vegetables in uncontaminated soil are increasingly difficult.

Sea level rise is no distant threat. Already some low lying islands are having to be evacuated, and coastlines are being squeezed all over the world, a process that scientists agree will continue for hundreds of years. So the main battle is to understand how much the sea will rise and how fast. How long is it before large areas of some of the most populated parts of the earth have to be abandoned to the sea?

As with so many other aspects of climate change there are uncertainties about how quickly the problem will grow into a large scale disaster. Imagine the upheaval for the countries involved. What happens to a nation's stability when the vast populations of low lying Bangladesh, the Mekong delta in Vietnam and the Nile delta in Egypt are forced to move. In each case these delta dwellers work the fertile farmlands which feed the nations concerned. Much of this farmland will be inundated. For these countries, and many others, the important question seems to be how quickly the waters are rising and what they can do about it.

For the layman the most obvious cause of sea level rise is the melting of glaciers and of the huge ice mountains of the Arctic and the Antarctic, now well documented across the world. The water runs downhill and meets the sea, like a dripping tap filling the bath.

Scientists, however, calculate that the main current cause of sea level rise is the warming of the oceans. Water expands as it warms. The atmosphere is being warmed by the man-made greenhouse effect, and because of the vastness and depth of the oceans it takes a long time for the water to absorb the heat. This is why scientists say that whatever we do now the oceans will go on rising for 300 years and possibly 1,000 years. We can, however, take action to slow the process down by keeping the temperature rises as low as possible, with the aim ultimately of bringing them down again. Working out how much oceans rise because of thermal expansion is complex because warm water expands much faster than cold. At 5°C a rise of 1°C causes an increase in volume of only one part in 10,000. However, at 25°C, a temperature found at the surface all across the tropics, a rise of 1°C causes a rise of three parts

in 10,000. If this 1°C rise happened across the top 100 metres of ocean in the tropics it would increase the depth by three centimetres.

Although logically, because water is liquid, sea level rise should be same everywhere, scientists do not believe this will happen. This depends partly on sea temperatures in different places and changes in ocean currents and air pressure, which can lead to dramatic rises and falls in sea levels. Some areas may see quite small rises and others very large. So far the computer models are still unable to predict where these changes might occur. About a third of the sea level rise in the last century is thought to have been caused by thermal expansion — and everywhere sea temperatures are rising. The effects are bound to be increasingly significant.

The next really big question is what is happening to the two biggest lumps of ice on earth, the Greenland and Antarctic ice caps. Some scientists at first believed that while some melting is occurring it should be discounted because a warmer world will produce more water vapour and therefore more snow. This would make the ice caps bigger not smaller. But the evidence is that this is not happening as much as scientists thought it would. Melting is speeding up faster than the snow can build up. However, the calculations of what this will mean for sea level rise are as difficult as for thermal expansion. Both ice caps are so enormous that they would not disappear completely for many centuries, and probably millennia. Only small fractions of them would have to melt to cause alarming sea level changes. Even a one metre rise in oceans would radically alter coastlines across the world and turn thousands of square miles of productive farmland into new seas or salt marsh.

Right: The Nile river and its delta seen from space. The Sinai peninsula and Red Sea are on the right. It shows what a tiny part of this vast land mass is occupied by human settlements and how Egypt is so heavily reliant on the Nile delta for growing the food that feeds the country. The construction of the Aswan Dam completed in 1970 cut off the silt that the annual flood brought to the Nile valley, which fertilised, replenished and extended the delta into the Mediterranean. Now sea level rise is gradually inundating the delta, reducing Egypt's ability to grow food.

There are a lot of uncertainties, and until recently many scientists believed that the melting of Greenland and the Antarctic was not a threat, at least not for another 100 years, but as with some other areas of climate science, as time has passed the news has continued to get more alarming.

Recent work for the Intergovernmental Panel on Climate Change (IPCC) shows that mankind has been over-optimistic in placing many of its largest settlements so close to current sea level assuming that this will not change. During the last ice age the sea was 120 metres below the present level because so much water was trapped in the ice caps. As the ice retreated between 15,000 and 6,000 years ago sea levels rose about 10 millimetres a year on average, before slowing right down and remaining more or less unchanged for the last 3,000 years.

But in the 20th century sea level rise began again. IPCC scientists were divided about how much and how quickly the oceans would warm and ice retreat in the face of higher global temperatures. Nearly 10 years ago they came up with a figure of between 11 centimetres and 77 centimetres in the 21st century, but have since revised it upwards. Information in 2005 from 177 tide gauge stations with a global range show that over the last 55 years sea level has increased around 1.7 millimetres a year. Over the last 10 years it speeded up again to more than two millimetres a year. But new research published in 2006 says even this is an underestimate. Eric Rignot, of NASA's Jet Propulsion Laboratory in Pasadena, California, and Pannir Kanagaratnam, of the University of Kansas, used satellite data to monitor the speeds of glaciers that drain the Greenland ice sheet over a ten-year period up to 2005. There were large increases, first in the south of

Greenland but after the year 2000 also in the north. Several of the largest glaciers have doubled in speed over 10 years and are now advancing at 14 kilometres (nine miles) a year. This has increased the amount of ice being dumped in the sea from 90 cubic kilometres in 1996 to 224 cubic kilometres in 2005. This is enough to cause an annual sea level rise round the world of 0.5 millimetre. The researchers estimate that global sea level rise has now increased to three millimetres a year.

Separate calculations show that increased glacier and ice cap melt could increase sea levels by two metres this century, according to the Sir Alistair Hardy Foundation, which supports a group of ocean scientists based at Southampton University. Add to that an addition of up to 40 centimetres for expansion of the seas because of warmer water and sea level rise will have potentially catastrophic consequences for low lying regions.

Although these new figures seem to be high, they are still low compared with the potential for sea level rise. Scientists have calculated how much ice, and therefore how much water, is contained in glaciers and various ice caps. If all the world's glaciers melted, and the vast majority are already in serious retreat, then sea level would rise half a metre. If the Greenland ice sheet melted it would rise an additional six to seven metres, and the west Antarctica ice sheet another eight metres. The West Antarctica ice sheet is vulnerable because it is grounded on rock well below sea level so there is potential for the sea to "float" it off, causing it to disintegrate. The speed with which this can happen was illustrated when the giant Larson B ice shelf in the Antarctic Peninsula disintegrated in a single summer. If all of Antarctica melted sea level rise would be 80

Left: The United Nations building in New York in 2005. The picture shows how vulnerable the city is to sea level rise. The basements of skyscrapers on the lower parts of Manhattan island are likely to be flooded before the end of the century, rendering worthless some of the most expensive real estate in the world.

metres — although that is said to be so far off as to be beyond worrying about.

What has changed in the last five years is new observations in Greenland and Antarctica. Temperatures have risen over Greenland and the Antarctic Peninsula far faster than elsewhere on the planet, sparking off much speedier melting of ice than was anticipated. Huge pieces of ice shelf have floated away from the Antarctic Peninsula, in total around 13,500 square kilometres in 50 years, leaving giant glaciers on the mountains with unhindered passage into the sea. In Greenland the ice cap is 2,100 metres (7,000ft) deep. It is already melting around the edges up to 1,000 metres above sea level in summer and exposing rock which has not been seen since before the last ice age, at least 20,000 years ago.

Professor Niels Reeh, from Denmark, who has been studying the Greenland ice for 20 years, has a different set of figures for this ice cap. He says ice losses in the four years between 1995 and 1999 were about 50 cubic kilometres of ice a year. This is sufficient to raise sea levels world wide by 0.13 millimetres a year. But he agrees that since then the rate of melting has increased as temperatures have continued to rise in the region and may already have reached 0.5 millimetres a year.

What is happening in Greenland is a perfect example of what constitutes dangerous climate change. The EU maximum limit of a rise of 2°C is a global average but the rise in Greenland is expected to be one and half times the average. At the same time scientists calculate that if the temperature in Greenland were to rise by 2.7°C then it would trigger irreversible melting of the ice cap. In other words a world average limit of 2°C would still be too high to avoid a long term

but unstoppable sea level rise of seven metres. In 2005 scientists meeting at the international climate change conference in Exeter revealed for the first time that in addition to the Antarctic Peninsula and Greenland, ice shelves in West Antarctica were melting too. This had been previously thought to be unlikely to happen for 100 years, but scientists using satellite calculations think 250 cubic kilometres of ice is being added to the sea each year, making a considerable contribution to the three millimetre annual increase. Work is still continuing to understand the forces causing the problem.

Apart from the increased temperature, what has speeded up these processes? In the case of Greenland and other glaciers which grind their way across rocks it appears that melt water can act as a lubricant. Melt water seeps down through the glacier to the bottom and allows the ice to slide much faster over the rock, often tripling the speed as it reaches the warmer melting zones or breaks off into the sea to form icebergs. In West Antarctica and elsewhere rising sea temperatures under ice shelves both melt and lubricate moving ice, again dramatically speeding up the process.

While the uncertainties persist, it is clear that this is another area where scientists have underestimated the speed of change, perhaps unwisely. It can lead to complacency in those whose responsibility it is to take action to combat the problem. As with global averages in temperatures, tiny increases in sea level rise each year do not seem to pose much threat. But while in a calm sea a two millimetre level change cannot even be noticed, it is different when magnified by storms, wind and tides. A combination of wind and tide is responsible for the serious flooding in Venice and St Petersburg. In eastern Britain and the Netherlands it is a

Right: People on wooden walkways in St Mark's Square flooded by high water in Venice, November 14th, 2001. At the beginning of the last century flooding occurred in the square, the lowest part of Venice, once or twice a year. Now the water comes over the banks of the Grand Canal and up through the paving stones and the mosaic floors of St Mark's cathedral more than 100 times a year. Venetians are abandoning the city as living conditions at street level become increasingly hazardous.

San Francisco

London

Miami

Rotterdam

Above and right: Venice may be the most famous example of a city facing inundation but many of the world's most populous and richest cities are on low lying coastal plains and are vulnerable to sea level rise. Although most coastlines have remained stable over the last 10,000 years, that is only because there has been no change in sea level. Without new sea defences to block the rising tides all these cities face waterfront flooding. London and Rotterdam already have barriers against the sea but need to build new ones to deal with tidal surges in the North Sea which can raise sea levels by as much as five metres. St Petersburg is also building its own barrier. Apart from the loss of life that a storm surge and flooding would bring to any of these cities, the economic consequences would also be vast. The world's major financial institutions and most expensive real estate, housing thousands of the world's highest paid workers, would be wrecked.

St Petersburg

Hong Kong

Bombay (Mumbai)

Singapore

northerly gale pushing the water down the North Sea to the bottleneck of the English Channel that is the danger. If this coincides with a high tide a massive warning network is activated. The surface of the sea can rise five metres above normal, potentially overtopping the defences which protect millions of people. The same dome of water occurs under the intense low pressure of a hurricane. Add to that the height of storm-driven waves and a high tide and the extra millimetres are multiplied many times into a more serious threat. The story of hurricane Katrina and low lying New Orleans could be repeated many times. Much of the southern United States, particularly Florida, is vulnerable to such storms and will increasingly be so.

But flooding above ground is not the only threat. Many countries use boreholes for fresh water. When too much fresh water is taken out, that is greater amounts drawn off than are replaced by rain, the levels of groundwater drop. On the coast this can have two disastrous effects. One is to cause the ground to sink and the second is to suck in salty water from under the sea. The result is both a loss of the water supply and a greater danger of flooding. Many places in the world, including parts of China, have sunk by as much as two or three metres, so much so they are now below sea level and have to be defended with giant banks.

The southeast of England has another unrelated factor which adds to the danger. During the last ice age the British Isles were only partially covered by thick ice. As a result the northern half of the land mass was pushed down into the earth's crust. Once the ice melted the north began to rise again, tilting the south and east downwards. The process still continues, so to add to the sea level rise danger, another one

millimetre a year of sinking needs to be factored into the calculations of London's flood defences.

So it is no longer a question of whether parts of New York, Hong Kong or Shanghai and dozens of other port cities will go under but when, unless they can construct robust sea defences. A one metre sea rise would put Shanghai, the great boom port city of China, one third under water, drowning the bottoms of the skyscrapers in the vast commercial centre. Hong Kong would face a similar fate. Even London, with the best sea defences in the world, is having to buy time by building yet more sophisticated and higher barriers to keep out the North Sea tides. Rotterdam, the biggest port in Europe, also has gates it can close when the sea threatens. Venice on the Adriatic, already flooded more than 100 days a year, and St Petersburg on the Gulf of Finland, which is sited on a former marsh, are building their own barriers to protect themselves.

The reason for most of these cities growing up where they are is exactly their proximity to the sea. It is no surprise, therefore, that both the most valuable and vulnerable parts of these commercial centres are closest to the waterfront. Their populations and businesses will have to be relocated inland. How soon this will begin to happen is one of the unanswered questions. At the moment, where they are able, for example on relatively narrow river estuaries, city authorities are putting their faith in building sea defences. For weeks and months at a time they can continue without any threat but at seasonal high tides or during storm surges they must have protection. These usually take the form of huge gates designed to hold back the waters when floods are forecast. But some places, like New York, are too exposed to defend and here local governments are realising that

Right: The Thames Barrier, London's bastion against a storm surge in the North Sea which would overwhelm billions of pounds' worth of property and flood parts of the underground railway system. Sea level rise means that even this sophisticated barrier is not going to be enough to save London and already the UK's Environment Agency is working on massive new defences. If nothing is done the barrier would have to be lowered on every tide by the end of this century to prevent daily flooding.

"Those who will suffer the most are the small island states, who neither have the space to relocate people and the infrastructure nor money to invest in protective measures... the challenges are daunting and time is running out. We are fast approaching the point of no return."

Maumoon Abdul Gayoom, president of the Maldives, 1998.

their best hope is to slow down sea level rise as much as possible. This means reducing their emissions, and campaigning for everyone else to do the same. This they hope will slow temperature rises and so keep sea level rise in check for as long as possible, although as we have already mentioned, some sea level rise is inevitable because of emissions over past decades.

But cities, with their large populations and expensive real estate, have less of a problem than many low lying island states. Many island states have no high land to retreat to and face partial or complete extinction. Ultimately entire populations will have to move to new countries and face the loss of their culture and national identity. Best known of these countries to the developed world is probably the Maldives. To many they are the idyllic palm-fringed holiday islands in the Indian Ocean, but to the people who live there, the Maldivians, they are a nation under siege from the sea.

The president, Maumoon Abdul Gayoom, has been a tireless campaigner on climate change since April 1987 when a storm caused severe damage to the islands. In October that year in a speech to the United Nations in New York he was the first head of state to alert the world to the danger that sea level rise posed to coral island states like his own. He has made countless speeches since and written a book, The Maldives: A Nation in Peril, and distributed it as widely as possible. Realising that the rise of seas is inevitable, although he still hopes it can be reversed in time, the country has a "safe island" policy, which will mean greater investment in sea walls, more solidly constructed buildings, elevated areas for vertical evacuation, and environment protection zones. The capital, Male, has a sea wall built with Japanese aid costing $63 million (£34 million).

John Bennett, from New York, a representative of the United Nations Environment Programme, who spent six months helping the islands recover from the 2004 tsunami, said: "From the highest levels of Maldives society to the average person in the street there seems to be an unusual degree of awareness about the threat that climate change poses to the country. For example, one night a taxi driver asked me if it was true that the entire country would be under water in fifteen years.

"The government has been developing and promoting a plan to build safer islands. But the destructive pattern of the tsunami waves showed that determining the characteristics of a safer island is not so simple. And persuading residents to leave their home islands, however attractive the offer, may prove challenging. Still, it is important for the government to be taking the threat of sea level rise seriously and exploring options.

"The Maldivians are very resourceful, but unless we in the rest of the world change course dramatically the country stands to pay a horrible price for our sins. How can we care, as we have, for the tsunami victims of today and yet turn an indifferent eye to the trouble that will befall their children and grandchildren because of sea level rise?"

But the people of the Maldives are not alone. They are one of 43 tiny island states from the Caribbean to the Pacific via the Mediterranean and the South China Sea which are most vulnerable to climate change. The first of these to go will be those like the Maldives which are composed of many low lying coral islands, and they are already vulnerable. Although the tsunami disaster which struck the Maldives at Christmas 2004 was caused by an earthquake,

Previous spread: The first glimpse of the Maldives most visitors get is of this exquisite turquoise-ringed spread of atolls set against the deep blue of the Indian Ocean. Depending on your point of view this picture could show you a group of islands and a nation extremely vulnerable to sea level rise or an idyllic paradise holiday destination.

Top and above left: The Maldives was known to travellers even in ancient times. Today many of the more developed islands, like Reethi Rah (top), offer five-star luxury and the holiday of a lifetime to those who can afford it.

Above right: The capital of the Maldives, Male, is home to a nation with a rich heritage dating back over two millennia. Today the government is battling with the problem of sea level rise by fortifying a number of islands against the sea as safe havens when storm surges threaten.

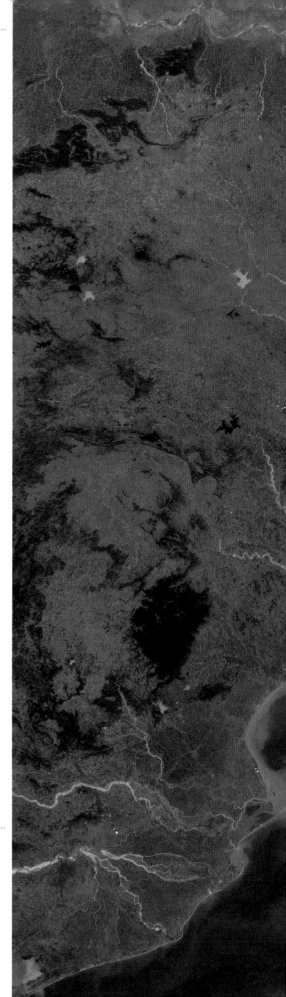

162

the damage caused by tidal waves was not dissimilar to that expected from tropical storms. Waves between one metre and five metres high reached the area and some islands were completely washed over. According to the United Nations Environment Programme 69 of the country's 199 low lying inhabited islands were damaged, 53 of them severely. Twenty were largely devastated and 14 had to be evacuated altogether. A third of the country's population, about 290,000 people, suffered losses and damage. These figures give some idea what a combination of sea level rise and tropical storms could do to the islands.

President Gayoom refused to talk to me about the possibility of evacuating his country, and is putting all his faith in adaptation.

"This is a scenario we do not have the means to prepare for, although as a country in the front line of climate change impacts we accept we would be the first hit and probably the worst hit. However, we may not be the only victims as time goes on. I believe the onus is on the entire international community to deliberate on how to deal with environmental refugees."

As the president said, his country will not be alone. The Marshall Islands, stretching almost 1,000 miles across the Pacific, like the Maldives, are likewise nowhere more than two metres above sea level and mostly less than half that. Among the nations threatened, just in the Pacific, are the Cook Islands, Fiji, Kiribati, Nauru, Papua New Guinea, the Solomon Islands, Tonga, Tuvalu, Vanuatu and Western Samoa. All will lose valuable land and some will disappear altogether. Already in Tuvalu some of the causeways connecting islands have been inundated. Families in the worst affected islets have been given sanctuary in New Zealand.

Some other islands are already untenable. In October 2005 the decision was made to evacuate the Carteret Islands, which are also in the south Pacific. For 20 years, the 2,000 islanders have fought against the ocean, building sea walls and trying to plant mangroves to hold back the sea and build up the soil. Each year the waves surged in, destroying vegetable gardens, washing away homes and poisoning freshwater supplies. The only food plants they were able to grow were planted in tubs above ground level. The islands are part of Papua New Guinea, and the government finally decided to move the islanders to the larger nearby Bougainville island, four hours' boat ride to the south-west.

In December 2005, at the climate talks in Montreal, it was announced that another small community living in the Pacific island chain of Vanuatu had been formally moved out of harm's way as a result of climate change. The villagers have been relocated higher into the interior of Tegua, one of the chain's northernmost provinces, after their coastal homes were repeatedly swamped by storm surges and waves. But in both these cases the islanders were at least lucky that they belonged to countries which internally still had somewhere to offer them to move to. When the time comes for other countries they will have to move everyone to the territory of another nation state. In a world of diminishing land area this means finding another country prepared to give up its own resources and provide another entire people somewhere to relocate. It is a problem the world has not yet faced.

Right: When the rivers of the Bangladesh delta are in flood following the monsoons they bring thousands of tonnes of silt with them from as far away as the Himalayas, adding offshore islands to this crowded country. Traditionally the poor occupy these new islands, build homes and farm them, risking inundation in subsequent cyclones. As sea level rise continues this natural expansion of Bangladesh will stop and be reversed as a large part of the delta disappears under the waters of the Bay of Bengal. This picture shows 600 square kilometres (240 square miles) of the area most at risk. Along these shores and similar low lying areas of India more than 100 million people will be displaced. No thought has been given to what may happen to these people and the tensions it may cause in what is already one of the most crowded parts of the world.

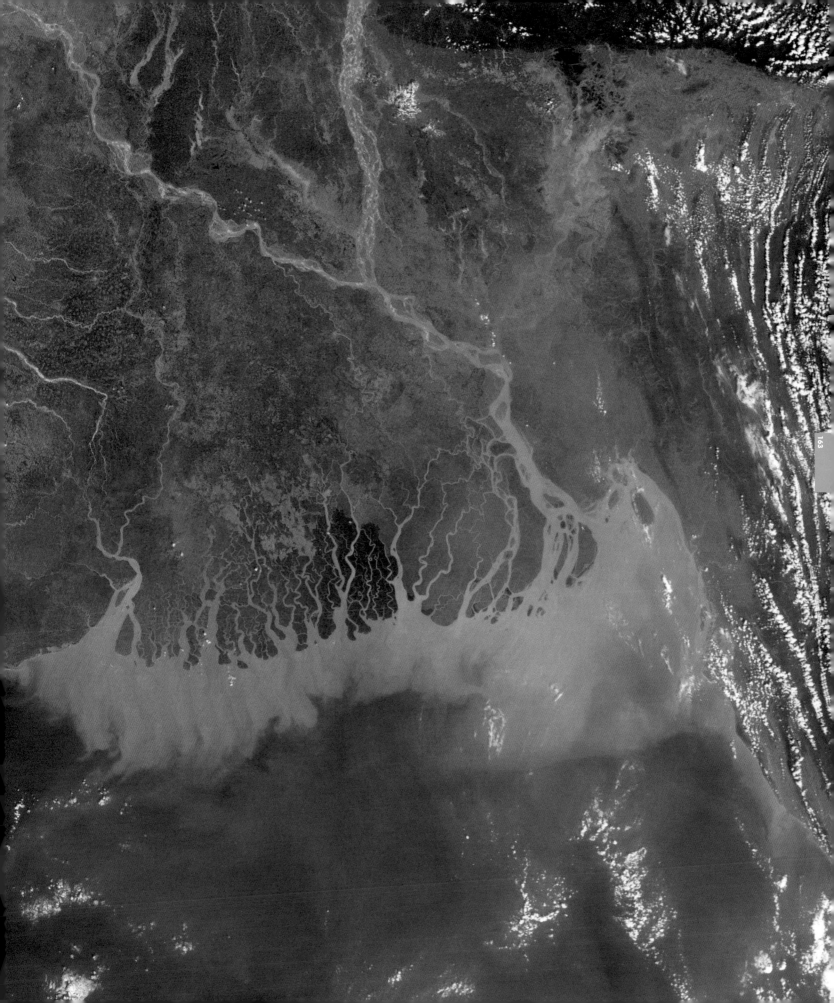

Sudden Cold?

With most debate about global warming centred on how quickly the earth is heating up it seems odd, not to say impossible, that some parts of the world might suddenly get much colder. But scientists have always said that climate change could involve some nasty surprises, and for western Europe being plunged into extreme cold could be one of them.

Analysis of ocean sediments shows that several times in the geological record average temperatures have dropped as much as 10˚C in as little as 10 years in Greenland as a result of dramatic changes in ocean currents.

These currents normally distribute heat round the world. The North Atlantic is kept at least 5˚C warmer than it would otherwise be because of the circulation of warm water from the tropics, which washes far up into the Arctic circle. Places as far north as Siberia and Alaska, which would be ice bound, have far milder winters as a result.

These ocean currents act like a conveyor belt. On the surface water moves north at about four miles an hour transporting water, warmed to 27˚C by the tropical sun, thousands of miles up the coast of Spain, France, Britain and Norway. It is like having a giant lukewarm radiator kept on all year round, gradually distributing its heat as it moves north. When the current reaches the Arctic the air cools the water sufficiently for it to sink and for ice to form on the surface. When the ice forms, the salt is left and makes the water underneath the ice denser, and heavier, than the fresh water around it. This cooler, saltier and denser water sinks to the ocean floor dragging in more warm water behind it. The current, equivalent in flow to 100 times that of the Amazon, is propelled back south, exactly like a conveyor belt, deep in the ocean, not resurfacing until it has travelled thousands of miles to the southern hemisphere. When it does resurface it starts to warm again to begin its journey back.

The strength and speed of the current is dependent on the sinking of the heavier salt water when it reaches the Arctic. If this did not happen the current, popularly known in Britain as the Gulf Stream, would be turned off. In fact the whole thermohaline circulation, as it is properly called, would cease to flow. This would have many dramatic effects on sea life because when it sinks the current carries vast reserves of oxygen to the ocean depths and keeps many strange life forms alive. For mankind the most obvious result would be the whole North Atlantic and the countries around it being plunged into cold. It would not just make life uncomfortable, lots of crops could no longer be grown because of the cold, and places like Iceland would probably become virtually uninhabitable.

This prospect has excited journalists in Europe and blockbuster film makers in Hollywood, and has alarmed some politicians sufficiently to provide funds to measure whether the Gulf Stream is about to stop.

So far the results are inconclusive. The current best guess from scientists from the climate research groups at the University of Illinois and other American universities is that there is a 50% chance of a total shutdown this century if emissions are not cut. The science is complex because there are a number of contributing factors to the loss of the current. They are: the warming of surface waters making them less dense; loss of the ice in the Arctic, which would dilute the salt in the surface water and prevent it sinking; an increase of fresh water into the Arctic from melting permafrost; more rainfall in the region; melting glaciers; increased river flow; and the loss of part of the Greenland ice cap, which would also dilute the salt. Some scientists believe that measuring devices across the Atlantic show that the Gulf Stream is already slowing down. Other computer models predict not a total shutdown but a reduction in the Gulf Stream of up to 50% by the end of this century. At the time of writing, however, the waters around Iceland and northern Norway are much warmer than normal. Why this is no one is sure, and it has happened in the past, only to cool down again. The warming may be due to increased hurricane activity in the southern North Atlantic, as discussed elsewhere, but scientists are still uncertain that they have understood all the factors involved in driving the ocean circulation.

Yet many researchers remain convinced that a sudden halt is possible, pointing to the fact that it has happened not once but several times in the past, although this was for natural reasons which are unlikely to occur again. The last time was at the end of the last ice age. When temperatures should have been rising in Europe they plunged 9˚C, and were cut all across the North Atlantic. If this happened in this century it would be economically and socially disastrous for western Europe and the eastern United States.

Shellfish on
Death Row

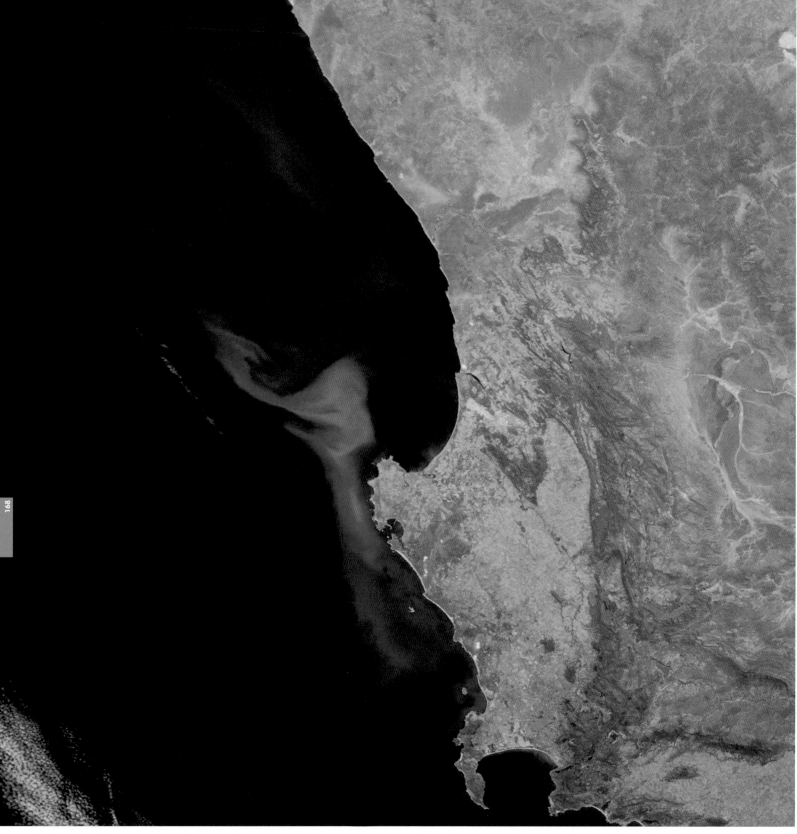

Previous spread: The oceans are under attack from climate change. One of the first major threats identified was coral bleaching. When the water gets too warm the tiny organisms that form the coral die, leaving the skeletons behind. It looks alright to the untutored eye, but parts of the Great Barrier Reef of Australia, shown here, which are normally swarming with life, are becoming ghost reefs of dead coral.

Above: A vivid turquoise streak colours the waters of the South Atlantic Ocean around Cape Columbine, South Africa. This sort of ocean colouring occurs when millions of phyto-plankton, microscopic marine plants, grow near the surface of the water. Such blooms are common in this region, where ocean currents sweep cold water from the ocean floor to the surface. The rising water carries nutrients from the sea floor, which nourish the tiny plants. The plants in turn feed fish and other marine life. As a consequence, regions that boast frequent phytoplankton blooms also tend to support a thriving ecosystem. Many tiny creatures, which form the start of the food chain, need the ocean water to be slightly alkaline so that they can extract minerals to build their skeletons. Increasing acidity in the sea caused by absorbing extra carbon dioxide threatens their survival.

Above: Large phytoplankton blooms are generally good news for all the animals higher up the food chain that eat them, but too much can also create dead zones in the ocean where no life can survive. As the tiny plants die and sink to the ocean floor, bacteria begin to break them down. When the plant matter is dense, bacteria can consume all of the oxygen in the water, leaving oxygen-free areas that cannot support life. This region off the east coast of Argentina is one of three known naturally occurring dead zones in the world. The other two are off the coasts of Peru and Oregon. Other dead zones, all man-made, occur near the mouths of rivers where agricultural runoff feeds phytoplankton. The largest in the world is where the Mississippi discharges into the Gulf of Mexico.

Scientists gave warning in 2005 of a newly discovered threat to mankind. They say it could wipe out coral and vast numbers of tiny ocean creatures on which many species of fish, whales and other sea life depend for survival.

Although the phenomenon is caused by excess carbon dioxide in the atmosphere, it is not a "global warming" problem, but a simple chemical reaction between air and sea. Carbon dioxide mixed with water produces carbonic acid, which is making the alkaline oceans more acidic. But for hundreds of thousands and probably millions of years, plankton, shellfish and corals have adapted to use the stable levels of calcium and carbon in the sea to make their shells.

This acid threatens to alter the balance of marine life. Chalk, perhaps most famously seen at the white cliffs of Dover, is made from the remains of trillions of these creatures, which still form a vital part of the present food chain.

So alarmed did marine scientists become about this new threat caused by the extra carbon dioxide in the atmosphere that in 2005 special briefings were held for British government departments. Carol Turley, head of science at Plymouth Marine Laboratory, warned them of a "potentially gigantic" problem for the world. "It is very urgent to warn people what is happening," she said. "Many of the species we rely on to eat, like cod, will disappear. The whole composition of life in the oceans will change."

Later research showed that while all the world's oceans would eventually be affected there were certain areas which were more vulnerable. These were the Southern Ocean around Antarctica and the tropics where corals form. This is mainly due to the way that ocean currents distribute the water with greater carbonic acid. In the shallow seas where corals form the impact of carbon dioxide is likely to be greater and the tiny organisms that create the reefs may no longer be able to do so because not enough will survive.

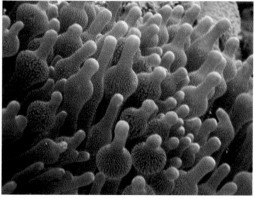

This page: These are the vibrant colours and teeming life associated with healthy coral reefs, the rainforests of the oceans where there are thousands of different species of animals and plants living together in a complex ecosystem. At the top is a type of Gorgonian coral with its colony of fish in a still healthy part of the Great Barrier Reef in Australia. Above and right are scenes from the Philippine coral community. These creatures are a bulb-tentacle sea anemone and a shoal of fish known as monos.

"Some scientists are saying that, in 35 years, all the coral reefs in the world could be dead – it could be less or more. Put it this way, my children may never get the opportunity to go snorkelling on a living reef. Certainly, my grandchildren won't."

Jerry Blackford, joint author of a paper on the acidity of the oceans, presented to the Exeter conference on climate change February 2nd, 2005.

Following spread: Contrast the scenes above with the sad task of Dr Peter Harrison, from Southern Cross University, New South Wales. He is standing surrounded by bleached coral on the Great Barrier Reef, still one of Australia's greatest tourist attractions and the largest coral reef on earth, trying to find signs of life. Warmer sea temperatures are killing the reef, wiping out the organisms that build the coral. Reefs can recover with time but not if there are repeated incidences of the water being too warm, as has happened in the last decade. Reefs are vital as fish breeding grounds and as a barrier to protect islands from storm surges.

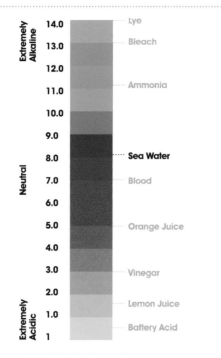

Extremely Alkaline	
14.0	Lye
13.0	Bleach
12.0	
11.0	Ammonia
10.0	
9.0	
Neutral 8.0	**Sea Water**
7.0	Blood
6.0	
5.0	Orange Juice
4.0	
3.0	Vinegar
2.0	
Extremely Acidic 1.0	Lemon Juice
1	Battery Acid

The sea has been slightly alkaline for millions of years because plants take carbon dioxide out of the water to grow. This process has kept the balance right for many shelled creatures to evolve. They take the chalky minerals out of the water to make their protective shells. In slightly acid water these shells would dissolve. The excess carbon dioxide in the atmosphere has meant the formation of carbonic acid, altering the balance between acid and alkaline.

Around a pH value of 7.0 is regarded as neutral – above that level, around 8.2, the water is alkaline and good for shellfish and coral formation. As it starts to drop towards 7.0 life gets more and more difficult for animals that rely on shells for survival.

The oceans' vital role in limiting CO_2 levels in the air will also have to be reassessed in the light of these findings. Plankton are as important as plants and trees in the take-up of carbon. Scientists estimate that about half the 800 billion tonnes of CO_2 put into the atmosphere by mankind since the start of the industrial revolution has been soaked up by the sea. Much of the carbon is fixed in the shells of creatures called coccolithophores, whose bodies have been making chalk layers for millions of years. They live on the ocean surface in trillions and when they die their shells sink to the bottom taking the carbon with them. They could not survive in a more acidic sea and their removal of carbon from the atmosphere would stop.

"These creatures are part of our survival bubble. The oceans give us a sustainable atmosphere by taking out the carbon dioxide. They're the lungs of the planet. People have not woken up to the potential impact their removal will cause," said Dr Turley.

The acidity of liquid is measured on the pH scale, from one to 14, with 7.0 being neutral, and the higher the number the more alkaline. The oceans have previously recorded an 8.2 pH reading, but this has now dropped to 8.1 and is continuing to fall.

Experiments show that even a small increase in acidity reduces the ability of shellfish and plankton to grow and causes a population fall. The loss of corals would seriously affect small Pacific islands, which owe their existence to the building abilities of these tiny creatures. Reefs also play an important part in the protection of coasts. The biggest fundamental problem is the effect on the food chain. Zooplankton, essential food for fish, could suffer increasing mortality rates and starfish, whelks and other shellfish,

Top: In the Red Sea the coral reefs teem with fish and other creatures. This seastar would probably not be affected by slightly more acid oceans but the coral on which it lives would be unable to thrive. Without its favoured habitat the seastar would soon be on the endangered list too.

Above: One of the potential winners from climate change. The jellyfish does not rely on the slightly alkaline sea water to build its body, and so would be unaffected by acid oceans. As other species die out there is a possibility that jellyfish will multiply hugely and take over the oceans.

"Those who recognise that the fortune and fate of humankind is inextricably tied to the state of the sea understand that trouble for the oceans means trouble for us."

Sylvia Earle, The Final Frontier in The Blue, dakini books, 1999.

Above: Humpback whales are not the largest of these giant mammals but are probably the best known. Their recorded songs or voices have been released on record and they are most frequently seen on film and by whale watchers because of their spect-acular antics and mating rituals on the sea surface. They eat across the world's oceans by sieving their food through giant plates in their mouths, which act in the same way as fishermen's nets. Many of these creatures that form the whales' main diet will disappear as the oceans become more acid, in turn threatening the whales' long-term survival.

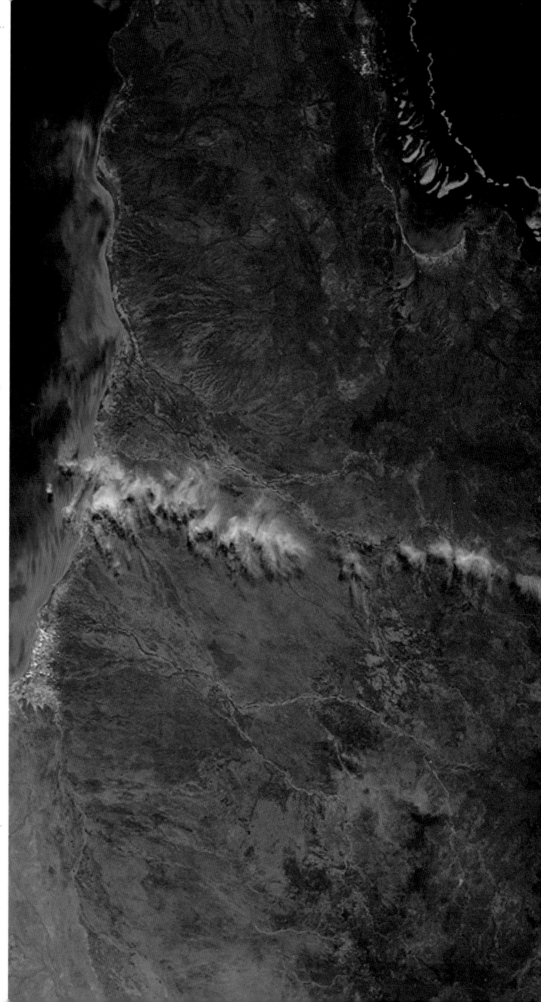

eaten by cod, might perish. This might lead to population explosions of other creatures, such as jellyfish, or crabs, shrimps and lobsters, which rely on chitin rather than calcium for their shells.

As has already been discussed in previous chapters, seas around the world are warming. This has obvious effects on sea ice, and at the North and South Poles on the species which have adapted to living on or underneath it. Not so well researched or understood are the changes in species that occur elsewhere. To the non-scientist it is surprising that a small change in water temperature makes a large difference to the species mix. Fishermen, of course, have long known this, simply because of the species they catch, but studying the changes scientific-ally is a great challenge.

Part of the reason for this is that man's other activities, apart from warming the oceans, have changed fish distribution already. Many of the world's favourite fish to eat, like cod and salmon, are already over-fished, so their disappearance from some areas has nothing to do with climate change.

However, in the North Sea, which has heavily studied shared fisheries, it is clear that climate change is making a difference to stocks. Tiny cod are dependent on certain types of plankton for food during the vital early stages of development when they live close to the surface. But the warming of the surface water has meant the replacement of their preferred food with less nutritious southern varieties. The result has been less breeding success for cod, and damage to a commercially important fishery. To thrive cod must move north.

Right: The magnificence of the Great Barrier Reef from the air disguises the fact that in many places the reef is dying. Partly to protect its fossil fuel industry Australia refused to sign the Kyoto Protocol, and supported the United States in its attempts to destroy the only legally binding agreement to reduce greenhouse gases. If Australia's efforts succeed and the world's politicians continue to do too little to combat global warming the country's other great industry, tourism, will suffer. Climate change will destroy the reef and ravage the country's other natural wonders, including its forests. Droughts and fires will also be common.

176

Fortunately for fishermen this has not meant empty seas. Sardines and anchovies are moving into the southern North Sea and red mullet and bass are extending their ranges northwards and westwards, as far north as Norway. In the south, Mediterranean and west African species are being found off Portugal. On the opposite side of the Atlantic off the Canadian coast some fish species have moved south. This appears to be in response to cooler and less salty waters caused by increasing melting of ice and glaciers.

One other related problem affecting oceans is so-called dead zones. These are usually where rivers or other pollution-heavy discharges go into the sea. Perhaps the best known, and largest, is in the Gulf of Mexico where the Mississippi river reaches the sea.
The main cause is excess nitrogen run-off from farm fertilisers, sewage and industrial pollutants. The nitrogen triggers blooms of microscopic algae known as phytoplankton. As the algae die and rot, they consume oxygen, suffocating everything including clams, lobsters, oysters and fish.

In 2004 a new report by the United Nations Environment Programme said that in the previous decade the number of dead zones had doubled to 150. They affected 27,000 square miles of ocean, an area the size of Ireland. As well as the Gulf of Mexico and Chesapeake Bay in the United States they were also spreading to the Baltic Sea, Black Sea, Adriatic Sea, Gulf of Thailand and Yellow Sea. They are also appearing off South America, Japan, Australia and New Zealand and were directly linked to human settlements, agricultural development and subsequent pollution.

Although it is a difficult problem, it is soluble. If we reduce the use of nitrogen fertilisers and prevent sewage being directly released into rivers and the sea, natural processes can cope with the nutrient load and prevent oxygen starvation. The dead zone problem could be made worse by global warming which could cause changed rainfall patterns and increased temperatures. For example UN scientists calculated that increased rain could cause a 20% extra flow into the sea from the Mississippi. Along with a 4°C rise in temperatures this would cut oxygen levels in the sea by between 30% and 60%. This would massively increase the area of the dead zone.

Left: The white cliffs of Dover, one of the most famous views of Britain, are visible across the Channel from France on a clear day. The whiter parts mark the newest cliff falls, because this landmark is already being eroded by sea level rise. The cliffs are made entirely of chalk, which are in turn made up of the skeletons of billions of tiny sea creatures that lived in the oceans millions of years ago. Their descendants still swim in the Channel and inhabit the world's oceans. As they die and fall to the ocean floor they create new layers of chalk. But these creatures will disappear with increasing ocean acidity because they will be unable to extract the minerals from the sea water to build their skeletons. At the same time, as with acid rain's action on many of man's ancient buildings, the cliffs will be further eaten away and undermined by the sea, as high tides wash against the base of the cliffs.

Extreme Events

Previous spread: Each tropical storm gets a name — most are rapidly forgotten but Katrina will live for a long time in the memory, particularly for the residents of New Orleans and along hundreds of miles of coastline along the Gulf of Mexico. Seen from space on August 28th, 2005 just before it struck the city, Katrina's swirling motion is producing winds of 257 kph (160 mph). The air pressure, another indicator of hurricane strength, is 902 millibars, making it the fourth lowest pressure ever recorded in an Atlantic storm. These two factors made Katrina a category 5 hurricane, the most violent kind. By the time the storm reached New Orleans it had decreased to a category 3 hurricane. Even so, the water surge was sufficient to overwhelm the levees protecting the city. The suburbs are below sea level and so the water subsequently had to be pumped out.

Above: Two men paddle in high water two days after hurricane Katrina devastated New Orleans and flooded large areas to a depth of 12 feet (3.5 metres) Many who stayed behind drowned when levees protecting the city gave way. Thousands were left homeless in the suburbs and beyond across Louisiana, Mississippi, Alabama and Florida.

183

Above: Scenes that shocked America and the rest of the world for days were those of the poor abandoned in the flooded city without food or the rule of law. It was four days before rescuers moved in to relieve thousands of desperate evacuees and brought them out of the city, shutting down two huge shelters that had become the scene of murder, rape and chaos. Here, in comparative luxury, a young hurricane Katrina survivor drinks from a bottle as others rest on camp beds on the floor of the Astrodome in Houston, Texas.

Hurricane Katrina, which wrecked New Orleans in 2005, focused world attention on the way extreme weather events can overwhelm man's defences. The city was warned, and apparently prepared, but was overtaken by catastrophe. Populated areas below sea level were found to have inadequate defences against such a storm.

Katrina was just one of a series of giant storms in the hurricane season which battered the coastlines of Central America and the southern United States. Records were broken with more storms over a long period than ever before. They appeared to be made more potent by the energy gained from the extra warmth in the oceans, particularly waters in the Gulf of Mexico. There was a total of 27 named tropical storms in the season, breaking the 1933 record of 21. Of these 27, 14 developed into hurricanes, again breaking the record of the most hurricanes in a season — the previous one being 12 in 1969.

One of the consequences of these storms was to stimulate a long-delayed debate in the American press and on radio. These news outlets had previously been dominated by those businesses and journalists who believe that talk of global warming is bunk. For the first time scientists and environmental groups who believe that climate change is a threat were given a sympathetic hearing. This did not prevent the White House intervening to try to silence the scientists. Jim Hansen, the director of NASA's Goddard Institute for Space Studies, mentioned elsewhere in this book, complained in January 2006 that the Bush administration had tried to censor him.

Until 2005 there had been considerable doubt in the scientific community that climate change was adding to the destructive force of extreme weather events. Problems arise because floods, droughts and windstorms have been causing havoc repeatedly throughout the centuries. They are recorded from the earliest times, whenever history was written down, but there were no precise measurements and the accuracy of the accounts obviously cannot be relied on. The computer models of the future, developed over the last 20 years, also contain estimates of actual conditions. They have predicted a

substantial increase in intense hurricanes but with a lower level of certainty than many of the other expected results of climate change. This doubt remains in the scientific community despite the simple fact that more heat in the atmosphere equals more energy, and therefore potentially more violent weather events. To have a hurricane, or what is known as a typhoon in the South China Sea or a tropical cyclone in the Indian Ocean, the temperature of the water has to be around 27°C (81°F) or more. This is why, as the oceans warm, there is a potential for the hurricane season to be longer and to extend the possible range of such storms north and south. Because of this, one of the predictions in climate models was that there was a potential for hurricanes in the South Atlantic for the first time. When the first recorded hurricane in the South Atlantic occurred in 2004 it caused a great stir in the scientific community, but little comment elsewhere. This was mainly because it hit a sparsely populated part of the coast and there were no casualties. But once again reality was catching up with computer predictions. However, some scientists still remain reluctant to say that man-made climate change has already led to more hurricanes.

The scientific reluctance to predict more dangerous storms, both north and south of the equator, was in contrast to the evidence from the world-wide insurance industry. Insurance companies repeatedly produced figures showing that their bills for weather-related insurance claims, and not just from hurricanes, were rising dramatically decade by decade. However, much of this is due to the increased value of assets along the coast.

In the days before Katrina struck, new data began to appear to remove the uncertainty about the violence of storms. Most predictions

"Katrina is the first sip, the first taste, of a bitter cup that will be proffered to us over and over again. It is up to us to tackle climate change and it does involve accepting that there is a legitimate role for government."

Al Gore, former US vice-president,
October 10th, 2005.

Above: Four scenes in the days following the terrible storm that swept through New Orleans at the end of August 2005. Although the strength of the storm was predicted, and the city was said to be well prepared, more than 1,000 died and there was no help for the thousands of poor people and some tourists left behind. Six days after the hurricane struck (top left), a victim is buried in a shallow grave, using bricks and sand. The woman, named Vera, had lain exposed to the elements. Bottom left, bodies were left uncollected on the streets while survivors, with what belongings they could move, tried to escape to safety. Even where the city was not flooded, the damage was considerable. Here (top right) a car was crushed by a falling wall, and bottom right, hurricane Katrina survivors are stacked five high in a Hercules aircraft as they are evacuated for medical treatment.

Dennis
July 10th

Emily
July 18th

Rita
September 21st

Wilma
October 23rd

Above: There were so many tropical storms and hurricanes in the South Atlantic in 2005 that meteorologists had run out of names by mid-October. Tradition dictates that successive storms are given names starting with successive letters of the alphabet. Although difficult letters like X are omitted there are normally more than enough names available for each storm season. In 2005, the Greek alphabet had to be pressed into use, and when the 23rd storm of the Atlantic hurricane season formed off the coast of Panama late on October 26th, it was dubbed Beta. The storm set a new record for the number of tropical storms to form in the Atlantic during a single year, a record that was broken with the formation of four more in November and December, which made it also the longest storm season on record. Scientists believe there is more energy in the system because the water temperature in the South Atlantic and the Gulf of Mexico was at record highs.

These storms, shown above with their names, caused damage over wide areas of the Caribbean and Central America where for ordinary home-owners it is no longer possible to get storm insurance. Homes destruction. Beta, for example, dumped 450 mm of rain (16 inches) in 24 hours in Nicaragua even though it was only a category 2 hurricane. These storms, shown above with their names, caused damage over wide areas of the Caribbean and Central America where for ordinary home-owners it is no longer possible to get storm insurance. Homes have to be built to be hurricane-proof before they can be insured. Hurricane-force winds from these storms can be felt up to 145 km (90 miles) from their centre and lesser but still storm-force winds reach 370 km (230 miles) away. The storm Dennis (top left) put clouds of rain over most of the southeastern United States well before the storm came ashore. In this image of Dennis taken on July 10th, the storm already covers all of Florida, Alabama, Mississippi, and parts of Louisiana. The economic cost of interruptions to normal life and damage by storms is measured in insurance claims, but the true total cost is impossible to calculate.

Irene
August 14th

Ophelia
September 11th

Beta
October 26th

Epsilon
November 29th

showed an increase in high-intensity hurricanes. Professor Kerry Emanuel, of the Massachusetts Institute of Technology, said that the destructive potential of tropical storms had doubled in the last 30 years. One of the reasons the danger had been underestimated was the failure to take into account the fact that oceans were warming to greater depths. Previously when a storm began the wind broke up the surface water, bringing up cold water from below. This cooling down drained the energy from the storm. But Professor Emanuel showed that with warmer waters at greater depths the cooling effect on mixing up the upper layers of water was reduced or lost, allowing the storms to continue to gather strength and last longer.

Almost as an illustration of this new science Katrina was in the same month spinning across the Gulf of Mexico towards New Orleans. It had left the Florida coast as a comparatively tame category 1 hurricane. Within a few hours it was gaining energy from the warm sea and reached the fiercest category 5 before weakening to a category 3 and scoring a direct hit on the vulnerable Mississippi coastline. Shortly afterwards a paper in the magazine Science showed that hurricanes in the most intense 4 and 5 categories had become twice as common over the last 35 years. The overall frequency of tropical storms had remained relatively constant since 1970, the article said, but the number of extreme events had gone up. Ocean surface temperatures have risen by an average of 0.5°C over the same period, almost certainly directly linked to global warming. In 2005 the waters in the Gulf of Mexico were a full 1°C warmer than the long term average.

By the end of the 2005 hurricane season, when the records were compiled, it was clear that it was not just the intensity of storms that had

increased. Significantly the season had started earlier. There had been more storms in June than ever before and the season lasted until December 6th, the latest date a hurricane has been recorded in the Atlantic. Hurricane Wilma, in October, also broke the record as the most intense hurricane ever recorded, producing 60 inches (more than 1.5 metres) of rain.

Insurance losses broke all records in the season — about $100 billion (more than £50 billion), of which $34 billion (£18 billion) was as a result of Katrina.

Despite these statistics, which go beyond anything suggested by the predictions of scientists from their own computer models, there remained argument among the experts, although new evidence in 2006 added to the weight of argument that the hurricanes were becoming more frequent and violent. Some still say that these extra storms are just a natural variation in the climate and not made worse by man-made global warming. This may seem extraordinary, but then to many it seems equally odd that New Orleans is being rebuilt in the same place. Despite the enormous goodwill towards the victims, particularly the poor, it cannot be sensible to rebuild the suburbs that are below sea level. The desire to avoid the centre of the city losing its real character and charm and becoming a theme park is understandable, but rebuilding the city as it was can only be short term. After all, whether the hurricanes are a result of natural variability or not, everyone agrees that another direct hit by a hurricane is a certainty sooner or later. Next time it could be worse. Katrina has already destroyed some of the off shore islands that help defend New Orleans from the storm surge and sea level is continuously rising.

Months after hurricane Katrina devastated large parts of New Orleans some suburbs remained in ruins. The storm struck in August 2005; in March 2006 members of the People Improving Communities through Organising (PICO) toured the region. They had chastised the US government in Washington for delaying bilions of dollars of aid for a region known for its historic churches. Top, Gloria Cooper of San Diego was still able to see this damage in the Lower Ninth ward of the city, and above, other members of PICO look at a damaged house in Gentilly neighbourhood.

Right: The scale of the damage to a once bustling city was only just becoming apparent two days after hurricane Katrina struck New Orleans. At this stage thousands of people were unaccounted for and rescuers had not reached many people still stranded in their wrecked homes. Many of the poor were trapped for up to a week by debris and floodwater, having been unable to join the pre-storm evacuation because of lack of transport.

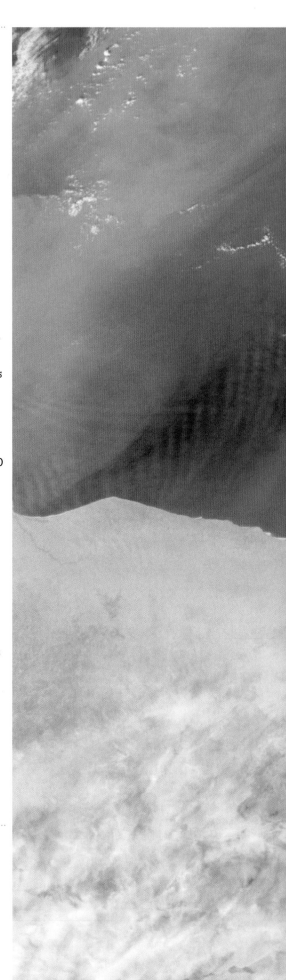

One other oddity of the season was a hurricane called Vince which began off the coast of Africa in October but instead of tracking west towards the Caribbean spun off northeast past Madeira. It eventually made landfall in Spain as a tropical depression. This is the first time this kind of storm has been recorded reaching the European mainland. Ironically the large quantities of rain deposited by Vince came as a relief to the Iberian peninsula because it had been suffering the worst drought in southern Europe since records began 60 years ago. Even orange trees died in the 42°C heat as irrigation water dried up. One of the long term predictions of climate change scientists has been the Sahara desert leaping the Mediterranean. The south of Spain, Portugal, France, all of Malta, Sicily, the toe of Italy, Greece and parts of countries bordering the Black Sea are all drying out. This is because of the misuse of water resources, for example for the tourist industry, and wasteful irrigation practices, as well as climate change.

Perhaps golf courses are the best, or worst, examples of this unreasonable use of water. Dutch engineers working in Spain on water conservation worked out that one 18-hole course can use as much water in a year as a town of 10,000 houses. Nine litres of water per square metre per day are needed to keep a golf fairway looking green in Spain. More than 40 new golf courses are planned for the Murcia region alone and as many again for Alicante and Almería. These are three of the driest areas of Spain.

Kofi Annan, the UN secretary-general, commented: "In southern Europe lands once green and rich in vegetation are turning barren and brown."

Despite what appears to be a developing battle between the competing interests for depleted water resources in southern Europe, these countries, with their wealth, are better placed than many on other continents. Even so, as a result of their plight these states drying out on the edge of the Mediterranean are all members of the UN Convention on Desertification and Drought. The convention has 179 members, and according to the UN 110 countries are already affected by desertification. The UN says the livelihood of one fifth of the world's population is threatened and an estimated 135 million people are at risk of being displaced. The worst affected continent is Africa. The UN estimates that by 2020 60 million people will leave the Sahelian region of north Africa if desertification is not halted. Nigeria, with the biggest population in Africa, is losing more than 350,000 hectares (nearly 1,500 square miles) of grazing and cropland to deserts each year. Kofi Annan says the spread of deserts, environmental degradation and poverty are all closely linked and must be tackled together.

But Africa is not the only area badly affected. The northeast of Asia has suffered severe dust storms in the last two decades and some settlements, schools and even airports have been buried in sand. Iran reported in 2002 that sand storms had buried 124 villages in the southern province of Sistan-Baluchistan. China has a tree-planting programme to try to halt desertification and even push back the frontiers of the sand dunes. Its target is to plant a billion new trees. Parts of North America are also drying out and have suffered dust storms, raising the spectre of the dust bowl of the 1930s. So far in the world as a whole the battle against deserts is being lost. Partly through natural desert spread, but also because of over-grazing, poor irrigation practices and

Right: A large, swirling mass of dust, visible in the top left portion of the image, is blowing from the Sahara into the Mediterranean Sea. The country on the left is Libya, while the Nile delta of Egypt is on the right. Dust storms occur when very strong winds carry sand from the great dunes of the Sahara. They are a naturally occurring phenomenon but are made worse by poor agricultural practices that contribute to soil erosion and continued desertification. The storms spread dust over vast areas of ocean, Europe and even the Amazon, depositing minerals as the dust falls. These can have a fertiliser effect in the sea, stimulating plankton growth and adding nutrients to aid tree growth in the forests. In the air they can also have an effect on rainfall. Scientists are still trying to understand their full impact, while on the ground governments and ordinary people are planting trees and creating other barriers with vegetation in an attempt to keep the sand in place.

"Africa is our greatest worry. Many countries are already in difficulties, and we see a pattern emerging. Southern Africa is definitely becoming drier."

Wulf Killmann, chair of the UN Food and Agriculture Organisation's climate change group, commenting on the fact that 34 countries including Ethiopia, Zimbabwe, Eritrea and Zambia were experiencing drought and food shortages, summer 2005.

deforestation, around six million hectares (nearly 25,000 square miles) of productive land are lost each year.

Each of the countries involved in the convention has developed a plan to take on the desert conditions in its own territory. So far the most successful schemes have involved local people, usually women, in planting suitable vegetation to hold back the sand. Large scale plans are often expensive failures because in harsh conditions trees and shrubs need individual attention and watering to survive long enough to take firm root.

As reported above, the UN has repeatedly made the link between poverty and land degradation. One of the big aims of the 2002 Earth Summit in Johannesburg was to improve the fresh water supply for the 1.1 billion people without it and provide sanitation for more than 2.4 billion. Lack of care in using water resources and keeping the organic content of soil are linked to the ability of people to grow their own food.

On a small scale, particularly in southern Africa, where these issues have been tackled together, it has been possible to reverse desertification and create thriving self-sufficient communities. These demonstration projects have yet to be adopted on a wide enough scale, but even if they were it may not be enough. The Hadley Centre for Climate Change predicts big increases in extreme and severe droughts across large areas of the world as a result of man-made climate change.

The problem of deserts has also been linked to the loss of forests. It has certainly been a major factor in land degradation. In the Amazon the destruction of once virgin forests is expected to lead directly to the creation of deserts. Over the

planet as a whole it is an amazing fact that already more than half the original forest cover has been destroyed, although much of this was in Europe and North America. Vast areas have disappeared in the tropics in the last three decades. It is possible to play with statistics to show how quickly tree cover is disappearing. In the 1990s the rate of forest destruction was 16.1 million hectares (62,200 square miles) a year. Put another way that is 4.2% of the world's virgin forests in a single decade, or in tabloid terms 33 football fields a minute, or each year an area the size of Portugal. Whichever way it is looked at this level of loss is clearly unsustainable.

As mentioned earlier, the fight to save the world's forests has been a long and unsuccessful one, although politically it appears that prospects are improving. Forests are vital, from the view of biodiversity as well as climate. More species inhabit the tropical forests than any other habitat on earth. Many of their secrets are still undiscovered, along with the potential benefits in food sources and drugs. They are all being lost before they are even discovered. Forests are worth saving for that reason alone, but trees are vital for many reasons to do with storing carbon, managing water resources and climate change. Trees capture carbon dioxide to grow, so the cutting down of the forests is releasing huge quantities back into the atmosphere. Some estimates are that around 25% of the extra carbon dioxide being released into the air by man comes from the destruction of forests. In addition, all over the world where forests have been cut down, particularly in mountainous areas, devastating floods have resulted. The Chinese government has ordered the reforestation of its mountainous regions to regulate the flow of its rivers, so trees are both holding back floods and halting spreading deserts. They are also used as an effective break

Left, top: Lionesses cross the dry Ewaso Ngiro riverbed in Kenya's Samburu game reserve in January, 2006. The lion, known as the "king of the beasts", is losing its habitat so fast that it may disappear from the wild. Increasing conflict between humans and the big game of Africa means that the territory where these animals can thrive is being squeezed.

Left, bottom: Wildebeest, ever alert for marauding lions, are framed against an African sunset, in what is a reminder of how majestic Africa's wildlife can be. Droughts, believed to be symptoms of global warming, and human pressure for land are cutting into the traditional migration area for the once huge herds.

196

Previous spread: The drought that hit the Amazon basin in 2005 stopped the flow of some of its tributaries, stranding river steamers, which are the main means of transport in a region without roads. Scientists fear that this drought may be the start of the drying out of the region that will lead to the destruction of large parts of the forest, which have been **described as the lungs of the world. The extra warmth of the sea in the Gulf of Mexico is believed to be part of the reason, along with the reduction in the size of the forest through fires and land clearance. Many of the bigger trees were so damaged by the drought that they are dying.**

against avalanches, and on the coast, protection again erosion, sea level rise and storm surges.

Trees effectively act as a giant sponge to soak up the rain and gradually release the water into streams. Even in Europe where winter and spring floods are becoming an annual feature, it has been found that trees planted on the edge of tributary streams can hold back and lessen the impact of pulses of flood water that are causing so much damage.

Trees also evaporate water through their leaves, creating more water vapour, and as a result forests release enough moisture to form new clouds and cause repeat rainfall further downwind. Forests are creators of their own weather in another sense. For one thing they have an uneven surface compared with grassland, and also they absorb more heat from the sun. This combination makes turbulent damp air rise, stimulating further rain. In well-wooded tropical regions this often leads to sharp showers in the afternoon after a day of sunshine. If grassland replaces forests there is a considerable reduction in rainfall.

Computer experiments show that if the Amazonian rainforest was cut down and replaced by grassland then rainfall in the region would fall by as much as 70%. There was speculation after the Amazon drought of 2005 that the loss of vast tracts of rainforest in the last 50 years was already having a serious effect on the region. By mid-summer the greatest river on earth had already dropped 1.8 metres below normal level to less than 15 metres deep. Rainfall was less than 65% of the July average. But as with all averages these figures disguise the extremes. Some tributaries, normally hundreds of metres across, dried out completely. Whole villages which relied on

fishing for trade and survival were abandoned and river boats, the only form of transport in some places, were grounded and left unmanned in mid-stream.

It was the worst drought anyone could remember. However, records in this region only go back 40 years, so not long enough to tell whether it could have happened before. But the drying of the rivers came after five years of low rainfall, a potentially disturbing trend. Environmental groups immediately blamed at least part of the problem on deforestation altering local rainfall patterns. Scientists also believe there may be a link between the drought and the extra warm waters in the southwest Atlantic. The warm sea drew the moist air from the Amazon and provided energy which contributed to the record hurricane season in the Caribbean. Either way, for the river communities of the Amazon one of the devastating problems was that, even where there was water, millions of fish died from lack of oxygen. Warm water has less oxygen. In some places rotting fish were floating across the whole surface of stagnant lakes and pools. Again time will tell whether this was a one-off extreme event or the beginning of a pattern.

Left and above: The lakes of the Amazon basin and the rivers are interconnected and teeming with wildlife in one of the hottest and steamiest regions of the earth. Drought is not unknown but very rare and on the scale of 2005 has catastrophic consequences for the wildlife and human population. Here a horse trots across the dry bed of Curuai lake. The fish congregated in ever smaller patches of water and streams until lack of water flow led to many drying out altogether. Even where some water remained, lack of oxygen meant that fish died in millions, leaving this fisherman with no harvest even when the rains returned.

Fiddling While the Globe Burns

Previous spread:
Demonstrators depicting
President George W Bush as
fiddling while the earth
burns are comparing him to
Emperor Nero who is said to
have done the same when
Rome was burning. The
environmental group Energy
Action staged their protest
outside the United Nations
Climate Change Conf-
erence in Montreal on

December 5th, 2005, calling
on the US to rejoin the Kyoto
Protocol. They argued that
the administration was out
of touch with its citizens.

Above: Trees are particularly
vulnerable to climate
change because they rely
on conditions remaining the
same for many years so that
they can grow, fruit and
flourish. They live a long
time and generally drop
seeds close by, and so also
migrate more slowly than
any other plant. Here in
Portugal a tree succumbs to
drought and dies.

Above: Drought is an
obvious danger to forests,
because dry trees make it
much easier for fires to
break out. Here a villager
tries an impossible task of
extinguishing a forest fire in
Abrantes, central Portugal,
on August 22nd, 2005. These
fires raged across the country for several weeks after a
two-year drought left the
whole region tinder dry.

Left: Kofi Annan, United Nations secretary general, speaks via a videotaped message during the ceremony to mark the entry into force on February 16th, 2005 of the Kyoto Protocol. The ceremony took place in Kyoto, the city where it was negotiated. Mr Annan, who was at the UN headquarters in New York, had been a vocal supporter of the protocol during the tortuous seven years of negotiations needed to reach this moment. It was celebrated by its backers as a lifeline for the planet but rejected as an economic straitjacket by the United States and Australia. On the podium are Wangari Maathai of Kenya, 2004 Nobel Peace Prize winner; Gínes González García, Argentine environment and health minister; Masao Nakayama, Micronesia's representative at the UN; and Joke Waller-Hunter, then executive secretary of the United Nations Framework Convention on Climate Change.

There were tears and congratulations in the corridors and in the conference hall at the end of the Montreal climate talks in December 2005. It was, as these moments often are, described as historic. But will historians record this mammoth conference as a success or a failure, or something in between?

If mankind is to survive the potential catastrophe he has created by changing the world's climate the foundations of a workable plan to adapt and cut emissions should surely have been laid in Montreal. An attempt was made, but will it be enough? The elated and sometimes tearful reaction to the agreement at the end of two weeks of bargaining was partly because of the relief of exhausted delegates after the conference had teetered on the edge of disaster several times. From the beginning the toughest task of the delegates of the 189 countries represented was to re-engage the American delegation in the process of combating climate change. For many, the fact that the administration of George W Bush was even prepared to talk in the future about how to reduce emissions was a victory. For others it was fiddling while the earth burned.

It is important to examine what happened at the conference for two reasons. The first is that the 1992 Climate Change Convention and its 1997 addition, the Kyoto Protocol, are the only credible international mechanisms for reducing greenhouse gases. As has been discussed earlier in this book in the section on the history of climate change, the 1992 convention's "ultimate objective" is the "stabilisation of greenhouse gas concentrations in the atmosphere which would prevent dangerous anthropogenic [man-made] interference with the climate system." In the 13 years between then and Montreal the science community had overwhelmingly reached the view that the human race was already responsible for dangerous interference with the climate, and was every day making the situation worse. It would appear then that the time for further discussions of what to do should have been over. The politicians at the conference, all of whom helped run countries that had signed

up to the convention, were therefore obliged to take action on policies and actions to solve the problem.

The Kyoto Protocol, the add-on to the original convention, provided the major mechanism for taking these steps by giving all industrial countries legally binding targets which must be met by 2012. Although all 36 industrial countries originally signed up to this deal, George W Bush repudiated America's target and his country's involvement as soon as he was elected. Only Australia followed his lead.

Three years before, in 1997 when the protocol was negotiated in Kyoto, President Bill Clinton had agreed to reduce America's greenhouse gas emissions by 7% by 2012. Bill Clinton was unable to get the US Senate to ratify the 1997 deal because many Americans believed it would make their industry uncompetitive with countries that had no restrictions on their greenhouse gas emissions. For its part the Australian government was trying to protect both its own energy-intensive industries and its large exports of coal.

To understand what happened at Montreal it is important to realise that these previous political manoeuvrings had created two conferences in one. The first involved all those countries that had signed the original convention, 189 at the start of Montreal, virtually all the world's nations. The second involved only the more select number who had also ratified the Kyoto Protocol. These came to a total of 156 of the 189. This second group, of course, included the 34 developed countries which had accepted legally binding greenhouse gas reduction targets, and excluded the two countries that had repudiated the agreement, the United States and Australia. The other members of the Kyoto club

Facing page, right: Still fighting political progress on climate change every step of the way, Paula Dobriansky, United States under secretary, democracy and global affairs, puts the White House point of view at a news conference during the UN climate change conference in Montreal on December 7th, 2005. Next to her is Harlan Watson, senior climate negotiator. The US did eventually agree to talk about taking action to curb climate change when the Kyoto agreement expires in 2012.

Above: Floods are nothing new in the United States, but as everywhere else the number of extreme weather events is growing. This is Conroe, Texas, George W Bush's home state and one of the country's driest, where the headquarters of America's oil industry is also located. Here in 1999, a grandfather carries his 11-month-old granddaughter to safety through rising flood waters while his wife holds the leash to their dog. They were evacuating their home.

"The more the climate is forced to change, the more likely it is to hit some unforeseen threshold that can trigger quite fast, surprising and perhaps unpleasant changes."

Richard Alley, Penn State University, January 2002.

Above: Rescuers save a one-year-old child from floods near Guayama, Puerto Rico, on September 10th, 1996, when hurricane Hortense struck Central America. Hortense brought torrential rain through the Lesser Antilles, Puerto Rico and the Dominican Republic. It caused 39 deaths and $158 million worth of damage.

are developing countries who are in support, including, of course, all the small island states that face extinction because of sea level rise.

In legal terms the Kyoto Protocol had only come into force on February 16th, 2005, after Russia had finally ratified. The Montreal meeting was therefore the first opportunity for all the participants to meet as a kind of club, and so the conference was officially the first Meeting of the Parties to the Kyoto Protocol, or in the conference jargon MOP 1. At the same time as the Kyoto partners were meeting to finalise the details of how the 1997 protocol would operate, a second parallel conference of all 189 countries that had ratified the original convention was taking place. This included the US and Australia, and was known as the 11th meeting of the signature nations of the 1992 Climate Change Convention. As described in an earlier chapter the first of these conferences had been held in Berlin in 1995. Under the terms of the treaty there has to be at least one conference of the parties, or COP, each year ever after. So the meeting was both the first Meeting of the Parties for the Kyoto Protocol, or MOP 1, and COP 11. Both had important business to transact.

This distinction between the two conferences may sound pedantic but is in fact politically vital because for the first time since the original convention came into force in 1992 the Americans were excluded from part of the discussions. The implementation of the Kyoto Protocol was a matter only for those nations that had ratified it. In Montreal the most powerful nation in the world was in effect an outsider, not a member of the club, and unable to influence its proceedings.

This change must be seen in the context of the previous seven years, when the Americans had

been free to take part in all discussions, because until 2005 the Kyoto Protocol had not been in legal force. For at least the last five of those years, since George W Bush had become president elect, the American delegations had done their best to undermine and destroy the process at each annual round of talks.

After first confidently expecting the Kyoto Protocol to collapse entirely in 2000 when America pulled out, and having been disappointed, US delegations had subsequently thrown every spanner they could into the works of the Kyoto Protocol. The fact that it still did not collapse and the rest of the world decided to proceed to tackle climate change without America was undoubtedly a constant irritant to the White House. It would be wrong to say the US was alone in its endeavour and did not have allies in this attempt at wrecking the protocol. Oil-rich Saudi Arabia and coal-rich Australia were constant in their support of the United States, creating delays and obstructions at every opportunity. Other countries, in difficulties with their own targets or seeking deals on technology or carbon trading, also covertly supported the US from time to time. They were happy to see progress slowed down but equally cheerful about the US getting the blame, taking care not to allow their domestic audience to see them playing an anti-climate hand.

There can be no doubt, however, that the US was the main player and ringleader and its fossil fuel industry the principal funder of anti-climate lobbyists and the greatest single influence on the White House. At Montreal, for the first time, the US could not play its traditional obstructive role. It was in effect out in the cold, outside the new club of nations. This had a significant psychological effect on proceedings, which moved much faster as a

Left: US President George W Bush walks hand-in-hand with Saudi Arabia's Crown Prince Abdullah among Texas bluebonnet wildflowers on his ranch in Crawford, Texas, April 25th, 2005. During his meeting with Crown Prince Abdullah, President Bush praised the kingdom's efforts to fight terrorism and sought the prince's help in bringing

down the price of oil. The two men have done their best to slow down international efforts to fight climate change. Saudi Arabia has the world's largest oil reserves and claims that combating climate change would damage its economy. At all climate talks since the Kyoto agreement was signed in 1997, the Saudi

Arabian delegates have demanded compensation from developed countries for loss of oil sales.

Above: Everywhere the US president goes he attracts demonstrators but an increasing number of them, as shown here during a visit to Brussels in 2001, are concerned about climate change. This was shortly after he had repudiated the Kyoto Protocol and said the US would never ratify it.

result. In 1997 when the Kyoto Protocol with its set of targets for industrial countries had been agreed in principle there had been a whole series of knotty problems to sort out in the small print. These were vital to the successful operation of the agreement. Among them were two methods of transferring technology to poorer countries, saving large quantities of emissions in the process.

The argument for these two schemes was simple. The climate did not care where the reductions in emissions were made as long as they were made. The protocol invented two ways technology transfer could take place between nations to the benefit of both countries involved; one was called Joint Implementation, the other a clean development mechanism.

The first relied on large sophisticated countries like Germany and Japan, which had already done much to make their power plants efficient, investing in other industrialised countries to bring them up to similar standards. Older plants in former Soviet bloc countries like the Ukraine, Romania or Bulgaria was incredibly polluting, but the governments concerned either did not have the money, or the technology, or both, to update the plants themselves. Under the treaty it was made possible for rich countries with good technology to pay for upgrades of equipment in less efficient ones and so claim "carbon credits" for cutting down greenhouse gases. Remember the argument was simply that the climate did not care where the savings were made as long as they were made.

The second process, the clean development mechanism, had the same motive, saving the climate, and was also a way of transferring technology to the developing world. The idea was that developing countries which wished to

undergo an industrial revolution themselves, to create wealth and employment, should not go down the same dirty route as Europe and North America had done. Instead new technologies like solar, wind power, and small scale hydropower would be installed in developing countries to provide electricity, leapfrogging over a century of dirty fossil fuel development straight into the 21st century. Those countries providing the technologies would be able to claim carbon credits that would count towards their own domestic targets of greenhouse gas reductions. This was a way of aiding the developing world without having to impose potentially unpopular draconian restrictions on voters at home. Some of the schemes were controversial and some technologies excluded. For example nuclear power, while a low carbon emitter, was not regarded as a clean technology. Large dams were also excluded and there was a great deal of debate about reforestation projects. How much credit should be given for planting trees to capture carbon, and changes in farming practices to prevent carbon being released from the soil in the first place, are a matter of scientific dispute, and as a result politically controversial.

One other difficult area that had to be settled was the penalties for failing to meet targets set down in the protocol. This is tricky because without sanctions on nation states, which would have to be noticeable, politically embarrassing and inconvenient, how can any legally binding international agreement be made to work? How could they be organised so that they did not also appear to infringe national sovereignty, a potentially touchy subject? Successful negotiations on sanctions had taken place at earlier meetings of Protocol parties but they remained to be brought into force in Montreal. The failure to reach the target will not become

Above and right: The old technologies and the new. Ratcliffe-on-Soar coal fired power station, near Nottingham, England, was built next to the coal fields, which once powered Britain's industrial revolution. The power station is one of Britain's biggest, but coal produces more carbon dioxide for each unit of electricity generated than

any other fuel. This global warming gas needs to be captured and stored underground if power stations like this one are to continue in operation without further damaging the planet. The lower picture shows one of the potential solutions, solar panels. These are in Pellworm, Germany, and produce power directly for use in the building on which

they stand without any carbon emissions. Top right is one of the oldest oilfields in the world, and one of the most polluted. The 100-year-old Iljitsh oil field in Azerbaijan on the Caspian Sea is a forest of rigs, some disused and some still operating but leaking, standing in lakes of oil more than a metre deep. This is in sharp contrast to the clean lines of

the offshore windfarm at Middelgrunden, Denmark, which, once built, produces carbon-free electricity.

"Those least able to cope and least responsible for greenhouse gases that cause global warming are most affected. Herein lies an enormous global ethical challenge."

Professor Jonathan Patz on behalf of the World Health Organisation, which estimated that 150,000 people a year already die and 5 million are made severely ill by climate change, November 17th, 2005.

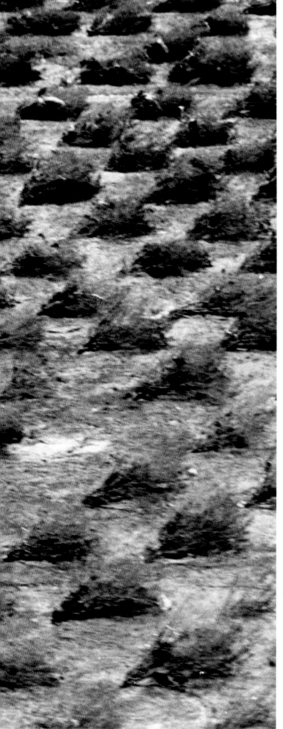

politically embarrassing until 2012, but by then the world might be far more concerned with climate change than it is now.

The working of the sanctions agreement will be assessed and applied by an expert panel. This is already set up and is monitoring progress towards targets. In 2012 it will ultimately judge whether each country has reached its commitment with a review of its final annual emissions inventory. If the panel concludes that a country has failed to meet the requirements the government concerned will be given 100 days to buy carbon credits or use any of the other bargaining or compliance mechanisms allowed for in the protocol. "If, at the end of this period, a party's emissions are still greater than its assigned amount, it must make up the difference in the second commitment period, plus a penalty of 30%," says the convention rule book. It must also develop within the following three months a "compliance action plan" to show how it is going to meet its target in the next commitment period. Although these can hardly be described as draconian measures they had taken years to negotiate, and there were fears before Montreal that there would be further hold-ups, objections and obstructions, as there had been at every previous meeting. Remarkably nothing substantial was raised. Without the Americans present observers remarked that there was "an optimistic atmosphere — a desire to get on with it", as one senior delegate put it. The MOP 1 meeting was concluded successfully and the Kyoto Protocol, at last fully in force after Montreal, became the first real test of man's ability to solve the climate problem. There remained at the end of the meeting seven more years for participants to put policies in place to reach the targets they had agreed to in 1997 in Kyoto. Some of these had been agreed by previous leaders from

different parties, but unlike the United States and Australia no other leader had gone back on their country's pledges. A few were already on target to meet the promised reductions but most were not, and accepted the need for a big new push to reduce emissions.

In most countries, when the crunch year of 2012 comes, the governments of 1997 and 2005 will be part of history and a new leader will have to face the blame and the penalty. Delegates realised this danger and the treaty has sensibly built in an annual reporting mechanism so that each government reports progress in reaching its targets. As each report comes in, it is painfully obvious which countries are failing. Under the treaty, when they are wide of their targets they have to offer a programme of how to get back on track. In 2005 many countries whose economies had been growing since 1990 were already struggling to meet targets.

But back to Montreal. The details of the Kyoto Protocol having been settled, the next big step was to conclude the second parallel part of the conference, COP 11. This, of course, included the United States. The principal task of this part of the meeting was to decide what was the next step — beyond the end of the Kyoto agreement in 2012 — to reduce emissions further. The treaty had built into it the requirement that by 2006 parties to the Climate Change Convention would be discussing what to do beyond 2012. Remembering that the Kyoto agreement overall only reduced greenhouse gas emissions from the industrialised world by 5.2% and scientists advised that cuts from the entire world needed to be between 60% and 80%, this was quite a tall order. If cuts of this magnitude are ever to be achieved then the whole world needs to be involved.

Left: The best intentions of governments can be laid waste by extreme weather events. Iran is suffering from desertification and the oil- and gas-rich country is attempting preventative measures. Here pine trees planted in a government reforestation plan lie flattened by floodwaters in the Minoodasht area of north-eastern Iran in August 2001.

Devastating storms in north-eastern Iran killed some 300 people, following what is thought to be the region's worst flooding for 200 years.

"It is clear from the work presented that the risks of climate change may well be greater than we thought. It is now plain that the emission of greenhouse gases, associated with industrialisation and economic growth from a world population that has increased six-fold in 200 years, is causing global warming at a rate that is unsustainable."

Tony Blair, UK prime minister, foreword to his department of environment report on global warming, January 30th, 2006.

Above: Some American politicians refuse to accept the term global warming and insist that it must be called climate change. In the sense that the extreme weather events can mean sudden cold as well as extra heat and bigger storms they are right. Here pedestrians make their way through snow and slush in New York in February 2006.

The thigh-high snow was said to be a record in the city and caused the cancellation of flights and delays to trains across the north-eastern United States.

Although it would be possible for the Kyoto parties, already adopting new technologies and measures to reduce emissions, to continue with the task, they could not alone reach a 60% reduction target for the planet. Without the United States, which produces 25% of the world's carbon dioxide on its own, no agreement could save the earth from uncontrollable climate change. In addition the developing nations, including the new industrial giant China, the ever-growing India, and rapidly advancing countries like South Africa, Brazil and Mexico, also needed to be brought into any agreement. They are all planning rapid industrial development, which implies increasing emissions of carbon dioxide and other potent industrial greenhouse gases, including nitrous oxide and methane. It was the "unfair" competition from these emerging economies that the United States used as an excuse for pulling out of the Kyoto Protocol in the first place. It had therefore become clear well before Montreal that all nations of the world must be brought into the process of reducing emissions and must accept their responsibilities to help save the climate. It does not take much imagination to see how difficult this is going to be but this is what the second part of the Montreal meeting was about: how to develop international agreements to reduce greenhouse gases after 2012. It was just talking about this idea that led the Montreal talks to the brink of disaster. There are some great issues of principle here that divide nations, and blocs of nations. First of all, as has been previously mentioned, it is the developing world's view that the old industrial nations, particularly Europe, Japan and the United States, are responsible for most historic greenhouse gas emissions. They therefore have the main responsibility for solving the problem. Through years of talks the constant refrain of the developing nations has

been that unless the developed countries, which have already grown prosperous by using the atmosphere as a dumping ground for their pollution, make a serious effort to reduce their emissions, then they, the developing nations, cannot be expected to take part. As a minimum they wanted to see these developed countries (including the United States) cut their emissions as part of the Kyoto Protocol agreement before discussing action in the next decade to 2020.

The second major stumbling block, also apparent before Montreal, is the refusal of the United States to consider in any future period the Kyoto formula of giving each country a target to reach and a timetable by which to achieve it. The majority of countries believe that only by legally binding agreements of this kind can they head off the danger of runaway climate change. The US rejects this notion. The position of the world's biggest single polluter is that new technology, either a single magic bullet, or a series of them, will be found to solve the problem. Against this background of fundamental disagreements it is hardly surprising that many people believed that the Montreal talks would collapse in disarray.

But there is a curious momentum to international meetings of this kind. In a world where everyone believes that there is a looming crisis with the climate, no politician, however powerful, wants to be remembered as the man who destroyed the talks aimed at saving the planet. Perhaps as important, in the run-up to Montreal it had become increasingly obvious that the threat was no longer some distant theory and that time was running out.

Another important change had gradually emerged over the previous year. China, the new industrial power, is seriously concerned about

Above: A rare sight anywhere, ice on an orange tree bearing fruit. Although, unlike other fruit, oranges need cool weather to ripen they do not need freezing temperatures. This scene, classed as an extreme weather event, was in Hillsborough County, Florida in January 2000.

Above: Tornadoes have been recorded for centuries and their destructive power is well documented. The question remains whether with more energy in the atmosphere, in the form of more heat producing more water vapour, the storms are getting worse. Modern technology is certainly making them easier to monitor. This rather grainy picture is the first Doppler radar picture of a tornado in the early stages of formation. It was captured in Oklahoma in 1973 and is in the National Severe Storms Laboratory (NSSL) collection. Scientists are still assessing whether tornadoes are getting larger, more frequent and causing more damage.

Above: This is one of the best images yet recorded of a waterspout, a type of tornado that occurs over water. Waterspouts are spinning columns of rising moist air that typically form over warm water. Waterspouts can be as dangerous as tornadoes and can feature wind speeds over 200 kph (125mph), a frightening prospect for a small boat that may be in their path. Many waterspouts form away from thunderstorms and even during relatively fair weather. This one was seen off the Florida Keys, arguably the hottest spot for waterspouts in the world. Hundreds form each year over the increasingly warm sea and in the humid atmosphere. Some people speculate that these water-spouts are responsible for the many unexplained losses of ships and aircraft recorded in the Bermuda Triangle region of the Atlantic Ocean.

Top: Everywhere the world's forests are under pressure. Usually it is the poorest people who move into the forest to hunt food and find a plentiful clean water supply. On the edge of Bach Ma national park, Vietnam, this family uses the forest as a resource for fuel; it needs about 20 bundles of wood a month for a family of six.

Above: Elsewhere virgin forest is being cut down for palm oil plantations and other cash crops, threatening the last stretches of forest occupied by rare animals like the orangutan. Here in Malaysia as a result of deforestation a three-year-old orangutan has had to be rescued and survives at Sepilok sanctuary.

Above: Often roads are a short cut to deforestation. Here in Anzaharibe-South, Madagascar, road construction creates a scar in virgin forest as a new area is opened for colonisation.

"If we don't do anything more than what we are doing now, greenhouse gas emissions will rise by 50% by 2030, whereas science tells us that we need to reduce them by at least 50%... our Montreal marathon is over, but we still have a long road ahead of us."

Stéphane Dion, Canada's environment minister, in his closing address as president of the UN climate change conference, Montreal, December 2005.

the effects of climate change. Its own environmental crises — power and water shortages, pollution and the spread of deserts — are threatening its prosperity, its ability to grow food and the welfare of its people.

China was not the only developing country where attitudes had changed. Some of the poorest countries in the world have realised that for them the climate change policies of the richest countries provide a potential opportunity. There is already provision in the Kyoto Protocol under the clean development mechanism for rich countries to claim credit by planting trees in developing countries. One of the little reported but significant developments at Montreal was the proposal by two poor countries, Papua New Guinea and Costa Rica, that they should receive financial compensation for keeping their forests standing. It provided a new slant on one of the great unresolved issues of the last 20 years; how to keep the world's remaining tropical forests from being cut down. The nations that contain the forests are among the poorest, and their single greatest asset is the timber in their forests. As far back as the Earth Summit of 1992 it was forcefully pointed out to the developed world that they could not dictate to others the fate of their forests. It was up to each country to decide what to do with its own natural resources. After all, Europe was once a giant forest but Europeans had long ago chopped it down. What right did Europe have to tell the rest of the planet how to manage its trees?

But in Montreal this argument took a new tack. Clearly, keeping a forest standing, rather than cashing in just once on the timber it contains, is a sensible long-term strategy. There is a lot of potential for forest products like nuts and the medicinal use of rare plants, as well as selective use of valuable timber. And a new development in these days is green tourism. Rich westerners will pay money simply to see such tropical forests and a chance to observe some of the rare animals they contain.

So ten countries, headed by Costa Rica and Papua New Guinea and styling themselves the Coalition for Rainforest Nations, came up with a surprise package in Montreal. They proposed that poor nations like themselves should be brought into the climate talks as partners. Their contribution to keeping the climate safe would be to maintain their forest cover. It would provide rich donor countries with an opportunity to gain or buy carbon credits by providing aid to keep the forests standing. The donors would also be able to use discoveries of new plants in the forests to develop drugs and for other uses. The credits would be granted on the basis of carbon dioxide saved from going into the atmosphere, and they would be used towards the donors' own domestic targets. As Carlos Manuel Rodriguez, Costa Rica's environment minister, put it: "If we do not recognise the value of these forests they will be cut."

Apart from being a rather good wheeze to save threatened forests this idea also has an interesting political dimension. It implies that some of the poorest developing countries, with no blame for causing climate change, are prepared to be brought into the process, and at least accept one important target — not to cut their forests down. It is the first time that developing countries have accepted such an idea.

Towards the end of the conference the new mood of developing countries became apparent when a maverick Russian suggestion failed to get any support. The Russians proposed, for reasons that never became clear, that

217

"There is no longer any serious doubt that climate change is real, accelerating and caused by human activities."

Bill Clinton, former US president, speaking to the climate conference in Montreal, Canada, December 9th, 2005.

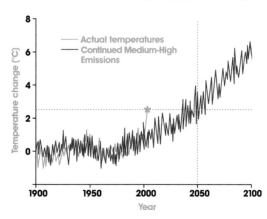

The heat wave in Europe that killed thousands of people in 2003 was unprecedented in recorded history. The scientists called it a once-in-1,000-year event. The asterisk marks how much above average the temperature was for more than two weeks in 2003. This graph shows the continuation of expected global warming after some attempt to reduce greenhouse gas emissions, but still leaving them at medium-high levels. By 2050 it is clear that the temperatures of 2003 will be normal in southern Europe. By the 2060s the exceptional heat of 2003 would be regarded as a cool summer. If the effect of pollution from traffic and industry, which blocks some of the sun's rays reaching the earth, was removed this kind of summer would even now be occurring more than once a decade on average.

Source: Hadley Centre for Climate Prediction and Research.

developing countries should not be asked to accept targets for cutting greenhouse gas emissions or timetables to do so. Observers expected that at least some developing countries would be in favour of such a free ride. None were, and the Russian proposal was withdrawn.

The new mood among developing countries is partly spurred on by the fact that the dangers of climate change are already obvious in many of them, and not just in the low lying island states. It is the poor that are always the worst affected.

None of that solves the problem of how to engage the United States in the process. The rejection of the idea of targets and timetables to cut greenhouse gas emissions, which the rest of the world sees as the way forward to deal with the crisis, leaves such a gulf between the two sides that it is hard to see how it can be bridged. To prove the point, towards the end of the Montreal conference, in a moment of high drama, the American delegation walked out. The talks were on the point of collapse and across the world it made headlines. One British newspaper summed up: "America alone against the rest of the world". But as I said, no one wants the blame for wrecking an important international conference. The White House appears to have been sensitive to the hostile reception that both domestic and world media gave the walkout. In the wake of hurricane Katrina, and the worst storm season on record, the American press, for the first time in recent history, was covering the climate talks in some detail. Within hours of the United States pulling out of the talks former president Bill Clinton spoke to the conference. It was a brilliant speech, and he did not pull his punches. His dismay at the US administration's action was widely reported across America.

So just as suddenly as they had walked out the US delegation was back in. Negotiations resumed and eventually, to the cheers and tears which began this chapter, a deal was done. But, beyond the obvious relief of those who had struggled through long days and nights to reach a deal, what had been achieved? Well essentially, every nation present including the United States had agreed to further talks about what to do to reduce greenhouse gas emissions after the Kyoto Protocol expires in 2012. As one of the detractors of the agreement put it, everyone had agreed to have talks about talks. Not much advance when the planet is said to be in such danger.

But let us go back to the thought that was uppermost in everyone's mind at the beginning of the talks. The major problem since President Bush was elected had been how to get America back into the international process for combating climate change. At the end of the Montreal conference, however small a victory it might seem, the US had agreed at least to talk about what to do beyond 2012 when the Kyoto Protocol expires, even if not an inch of ground had been conceded on what that might entail. In fact in the final conference statement, the Bush administration insisted on a get-out clause, which specifically excluded the United States being tied to any targets to reduce emissions or timetables in which they must be achieved.

Within a few days of the talks ending the gloss put on the agreement by politicians and some relieved environment groups was already being questioned. The British magazine New Scientist, read worldwide by scientists who keep up with the issue of climate change, summed up: "In the cold light of day we have to ask what exactly was achieved. The answer looks like little more than an agreement to carry on talking — and

Right: Just after the US delegates walked out of the Montreal climate change talks in December 2005, the former US president Bill Clinton rose to speak to the delegates. Because the talks were held just across the border many more US journalists than would normally cover the talks were there and the powerful Clinton speech criticising his successor in the White House was widely reported. He said the Bush administration was "flat wrong" to reject the Kyoto accord and said cutting greenhouse gases was good for business and the planet. Within hours the US negotiators returned and the conference was saved from collapse.

Above: Germany has always taken a lead in Europe on environment issues and the new Chancellor Angela Merkel has embraced the drive for renewable energy and unleashed a massive programme for energy efficiency in older housing and offices. Here the environment minister, Sigmar Gabriel (left), Chancellor Angela Merkel, and the minister of economics, Michael Glos (right), address a news conference after an energy summit in Berlin on April 3rd, 2006. German firms that harness renewable energy sources like the sun and wind to generate power said they planned massive investments in the coming years and want preferential treatment.

Above: Al Gore, who appeared to the outside world to fade away after conceding defeat to George W Bush in the battle to succeed Bill Clinton as president of the United States, re-emerged as a firebrand campaigner for climate change. The former US vice-president had represented his government at Kyoto and accepted a 7% cut in emissions for his country but President Bush pulled out. Al Gore toured the world in 2006 to alert people to what he describes as imminent disaster. Here he speaks at the 2005 Milken Institute global conference in Beverly Hills with a passion observers say he never possessed during the presidential elections.

even that is hedged in places by promises to talk about very little that is meaningful." It pointed out that while politicians talked "every square metre of the planet's surface is absorbing about one watt more heat than it can release into space. That may be only slightly more than the power of a Christmas tree light bulb. But it matters." The magazine said that in recent months it had reported that "we may face runaway melting of Arctic sea ice, a shutdown of global ocean circulation systems, massive methane releases from melting permafrost, stronger hurricanes and 'megadroughts' from northern China to the American west. These are not abstract outputs from computer models but things that are starting to happen."

New Scientist concludes that the Montreal agreement had done too little to quieten the fears repeatedly expressed by senior scientists including George W Bush's top climate modeller, Jim Hansen. The director of NASA's Goddard Institute for Space Studies had predicted, even as the politicians in Montreal were meeting to fudge the final communiqué, that at most the human race had 10 years to make drastic cuts in emissions or face the possibility of runaway climate change.

So despite the Montreal talks being hailed as a success it is clear that the gap between the scientists' recommendations for urgent action and the politicians' lazy progress continues to widen. Optimists say that at least talks will continue and countries like China and Brazil will explore what measures and targets they can contemplate, having previously refused even to consider them. Meanwhile the US was back in the process, and although the Bush administration is regarded as a lost cause, there were only three years before he was due to leave office. Already, they say, the mood in the

US following the 2005 hurricane season is changing. Many cities and states, which were already taking action on climate change, have reinforced their efforts. There is also a tide of opinion in industry that wants to seize new opportunities and begin the manufacture of new products that will cater for consumer demand for cars and greater electricity supply, but at the same time does not harm the climate. There is a feeling in some sectors that the US is being left behind in this new industrial boom and that the old fossil fuel industries and car manufacturers are holding the country back.

Optimists also point to the fact that there are now powerful advocates for action on climate change in both Republican and Democrat camps. There is widespread belief in America that no new presidential candidate would be electable with the same hard line of no action on climate. Another event like hurricane Katrina in election year could scupper any chance of such a candidate reaching the White House, or so it is claimed.

But a new American president would not be in the White House until 2009. Climate talks have always taken years to reach conclusions. Kyoto took three years to negotiate and eight more years to come into legal force. The clock is already ticking on Jim Hansen's 10-year timetable. The battle to save the planet clearly cannot be left to politicians alone. Something else has to happen which might save the planet. What might it be, and what can the rest of us do?

Above: Al Gore's message is partly that if politicians are slow to act then the pressure must come from the public. Citizens all over the world are increasingly demanding action as they see their way of life and their future threatened by climate change. Greenpeace, which has become a multinational organisation in its own right, campaigns in many countries for action. This picture shows Greenpeace activists displaying a banner celebrating the enforcement of the Kyoto Protocol in Kyoto, western Japan, on the day the treaty came into force, February 16th, 2005. But they are not alone; hundreds of environment groups, and many organisations concerned about human welfare and poverty, see climate change as a threat and are putting increasing pressure on governments to take the action needed to combat it.

Don't Mention Population

One of the great unmentionables in the climate debate has been population. At the 1992 Earth Summit it caused an enormous political row behind the scenes because the Vatican refused to discuss the issue of birth control.

But the fact that the population of the planet tripled in the last century to 6 billion cannot be ignored. Even with a currently falling birth rate in almost every country the number of people on earth is expected on current trends to rise another 2 billion by the middle of this century before stabilising.

Predictions made in the 1960s that this amazing rise in the number of people would lead to starvation have not materialised. The population rise was matched by what was called the green revolution, the use of fertilisers which increased yields. There were also large increases in the use of irrigation and much farmland was developed from former forest.

There are now those who think that the predictions of 40 years ago were merely delayed by these factors, and food shortages will become normal. They give as an example the appalling drought in East Africa in 2006, which put millions at risk. The number of malnourished people in the world is increasing. It is close to 900 million in 2006.

Lester Brown, of the Earth Policy Institute in Washington, believes a projected increase to 9.1 billion people by 2050 is "highly unlikely, considering the deterioration in life-support systems now under way in much of the world". He points out that already 42 countries have populations that are stable or falling as a result of falling birth rates. But while extra population obviously uses more resources, and puts pressure on water and wood supplies particularly, it is only part of the problem. The consumption patterns of all these people are also key. Perhaps the most telling and most often quoted example of this is the United States. This super-rich country produces 25% of the world's carbon dioxide with under 5% of the population. But almost as important in the debate is diet. It takes 10 times as much water to produce beef for the North American market as it does to produce grain for the largely vegetarian diet of Indians; it also takes far more land. Cattle and other animals bred for meat produce large quantities of methane, even more than that from rice paddies. In terms of consumption and production of greenhouse gas emissions, 100 Indians do less damage to the climate than 10 Americans.

As with most of the problems discussed in this book there are no easy solutions to this difficult and contentious issue. Religious beliefs and human rights issues clash, as well as the efforts to relieve poverty and preserve resources. China has tried the controversial approach of keeping population down by limiting families to one child each. Most of the rest of the world has tried educating women, and improving access to health care, as a way of reducing birth rate. Clean water and sanitation, which cuts the needless and appalling 5 million death toll caused by dirty water each year, mostly among poor children under five, can also have the effect of cutting family size. People no longer believe they need to have many children to ensure that some survive. Education and health care are startlingly successful means of stabilising population growth in some states in India (Kerala is an example). They also often help greatly in reducing poverty. This is in stark contrast to other regions in India, which have failed to help the poor in this way.

In the developed world population is already stable, and in many cases falling, particularly in countries like Germany, Luxembourg and Russia. However, carbon dioxide emissions continued to rise in all three countries in the first five years of this century despite efforts by governments to reduce them. This shows that the link between population and climate change is not so important as the lifestyle of the people involved. The rise of China as an industrial power, and the aspirations of its people to switch from bicycles to cars is another well-quoted example. But the reality is worse. Across the world around 80% of the population will soon be in developing countries and 20% in the developed world. Those 80% of ordinary people regard development as a greater priority than avoiding climate change, that is if they think about it as a choice at all. The International Energy Agency predicted in its 2005 report that soaring energy demand would increase carbon dioxide emissions by 52% by 2030 on current trends. Initiatives on cleaner energy could reduce demand but even then there would be an increase of 30%. This is still too much, the agency says. The task then is not just how to slow down population growth and feed all those extra mouths. The second need is to develop technologies so that these new con-sumers, and the rest of the population, can live in the style they want without wrecking the planet in the process.

Peak Oil
and China

225

Previous spread: The sheer scale of everything in China, from the speed with which its economy is growing to spread of its already vast deserts, is hard to take in. By China's standards this is a small-scale event. The middle school pupils of a city called Guangszhou, in Guangdong province, attend a fair for high schools looking to recruit students before they take a high school entrance exam. More than 60,000 people took part in this consultation on May 15th, 2004.

Above: This is perhaps how outsiders think of China. Irrigated rice terraces cultivated for centuries to feed its ever-growing population, now more than 1.3 billion. The effects of climate change and scarcity of water are making rice more difficult to grow because it is a very thirsty crop.

Above: In China, regions
dominated by mountains
and desert are so harsh that
thousands of square miles
are sparsely populated.
Here the Huang He or Yellow
river is at the beginning of
its long journey to the sea.
Its waters are used for drinking, irrigation and industry
to such an extent that only a
tiny fraction of its mighty
flow ever reaches the sea.

Probably the best hope for the world is that the oil begins to run out. In my view the quicker the better, although others disagree, because they fear that sky-high oil prices will cause an economic slump, preventing the world having the money to adapt to climate change.

In any event there are signs that oil demand is beginning to exceed supply, even though politicians continue as if it were not so. For example, China's long term plan to 2050 includes having half the country's projected 1.5 billion population owning their own car and being able to afford overseas travel. The plan, published in February 2006, included lifting another 80 million Chinese out of poverty and moving 500 million peasants to cities to work in factories and produce more consumer goods. This vision, which includes continuing the country's astonishing 8% annual growth rate, is impossible. If China grew at that pace its demand for raw materials would strip the world of natural resources. Already commodity prices, for metals like copper, have increased dramatically. But the worst shortages would be in water and oil.

As far as water is concerned, China and large parts of the rest of the world already have a problem verging on a crisis. Twenty-five years before China's grand plan is supposed to come to fruition 3 billion people in the world will be facing water shortages. At least 500 million of them will be in China, a country already over-using its water resources. To grow one kilo of rice uses up to 5,000 litres of water, so China is diverting its water to factories and cities to produce the wealth so that it can import food rather than grow grain itself. Even so there is not enough water in dams, rivers and replenishable near-surface aquifers to go round. An estimated 100 million Chinese are already relying on non-replaceable fossil water from deep underground to water crops. It is being pumped to satisfy short term demand. But while water is essential both to prosperity and to life it is a local problem. If China misuses its water resources, or simply runs out, it is China that suffers most and has to solve the problem. Both

Beijing and Shanghai are short of water. Shanghai is sinking as excessive use is made of groundwater. The Chinese government has fantastic plans to transfer water from the Yangtze river in the south to replenish the dried up Yellow river in the north. The cost is £40 billion and with that price tag China may be able to pipe water across country or even from neighbouring states but it cannot buy water on the world market. It is not yet a financially valuable commodity like oil, which can be traded across the world. The Chinese, who are also aspiring to eat more meat, are relying on importing food to meet their people's wish to mimic western diet. Whether this is possible remains to be seen.

Oil, on the other hand, is central to the continuous and unsustainable growth in the international economy. Its continued supply and availability has been part of the inbuilt belief of politicians and businesses alike, that the future always has the promise of increasing world trade and affluence. Part of this assumption, bar a hiccup or two, is that the oil supply is always enough to keep the price low. This assumption allows Chinese economists to believe that half the country's population in 2050 will be driving cars, and that the country's economic power will give it the cash to buy the necessary oil on the world market to power them. Even optimistic Americans no longer believe that. The Chinese grand plan was launched in the same month as President George W Bush, in his state of the union address, began a campaign to wean his country off its dependence on oil. President Bush did not spell out exactly what this meant to America but he was acknowledging for the first time that the demand for oil for the world's automobile fleet would soon outstrip supply. Even the US realises that the days of cheap oil are over, and that there will soon not be enough

"In little more than a decade, China has changed from a net exporter of oil into the world's second-largest importer, trailing only the United States."

Peter Goodman, Washington Post, July 13th, 2005.

229

Above: This is modern China, as the government likes to portray it: new factories and industrialisation to fuel an export boom that will make the country the world's next economic superpower. Top left is Kwei Lin silk textile factory. Outside China it is the country's thirst for metals and particularly steel that has created a stir, not least in the rising price of scrap metal, much of which ends up providing raw material for the country's construction boom and new industries like car manufacturing. China was the world's top steel producer in 2005 and the country is the world's fastest growing market for cars. In Shanghai, top right, a Chinese worker operates a machine for hot-rolled steel at Bao steel factory. Bottom left, a worker at the Anshan steel factory in northeastern China and bottom right, some of the steel bars end up at the construction site for Wukesong indoor Olympic stadium in Beijing.

Top: This is the other side of China, the battle against desertification and environmental disaster. Farmer Feng Yongcun, aged 74, gazes out at a 10 km (6.2 miles) long advance column of the Gobi desert looming over his village in Longbaoshan. The sand dune was a curiosity 6 km (3.7 miles) from the cornfields of his village north of Beijing when he was a young man but it now looks set to overwhelm his home. The "flying desert" now regularly darkens the skies of the capital Beijing and sand rains down on the city.

Above: The country has always had deserts, but now they are rapidly getting larger and semi-arid regions are being overtaken as here at the Jiangxi nature reserve. Staff survey the encroaching sand dunes at Poyang lake, Jiangxi province.

Right: A Chinese worker washes dust off the bushes and pavements in the centre of Beijing in an attempt to clean up the city during the periodic dust storms caused by strong winds from the Gobi desert in the north. People with breathing difficulties are advised to stay indoors for their own safety during these events.

to go round. How much oil there is and how long it will last has long been debated without any firm conclusion being reached. But that is not what matters most. The important question is, and always has been, how soon demand exceeds supply. Optimists, including the US government, have always said this will be another 20 years, a figure that remains the same with the passing of the years. The so-called tipping point (as far as oil is concerned) is when prices will begin to rise and never come down again because demand exceeds supply. This is always going to be years away. Pessimists, who also call themselves realists, think the tipping point may have already been reached, or will have passed before 2010. At the time of writing the price of oil has just closed in New York at a record high of $77 a barrel. It is too early to say whether this is the first sign that the tipping point has been reached. In my view the moment is close and at that point the price surge will continue, some predicting oil will soon cost more than $100 a barrel. Most important, the price will never be low again.

For the world economy many believe this will be a show-stopper. Not that higher oil prices necessarily stop growth, although they will cause it to stutter, but they will turn upside down the accepted theory of the way the world works. The current economic model is that demand stimulates supply, and that natural resources can always be found to fulfil that need. To some extent that does work with oil. Smaller marginal fields that were too expensive to exploit in the era when oil was $20 a barrel become very attractive when the price is $100. The problem is that the demand for oil is continuously rising and the number of discoveries of new fields is not keeping pace. New oil fields, when they are found, are no longer the easy and cheap-to-pump reserves of

the Middle East. Vast reservoirs of oil found just below the desert surface provided 50 years' worth of cheap oil. In the 21st century finding and then exploiting new fields involves ever more expensive deep drilling. These fields are either offshore, or in obscure and inhospitable regions, a long way from markets. Some are in shale deposits, which up to now have been too expensive to exploit. Ever longer pipelines or supply routes make the oil expensive to deliver to customers and vulnerable to attack. Even without the possible disruption caused by the continuing instability in the oil producing regions of the world, satisfying demand is becoming a more difficult and expensive struggle. The average price per barrel can only ever be upward.

Lester Brown, the previously quoted founder and president of the Earth Policy Institute in Washington, has spent many years studying the stresses on the world's environment, and particularly the effect of China's economic development on the world.

In a speech to the Organisation for Economic Co-operation and Development in Paris in February 2006 he summed up: "Many earlier civilisations at some point found themselves on an economic path that was environmentally unsustainable. Some understood what was happening and were able to make the needed adjustments and survive, even flourish. Others either did not understand the gravity of their situation or, if they did, could not adjust in time. They collapsed.

"Our global civilisation today is also on an economic path that is environmentally unsustainable, a path that is leading us toward economic decline and collapse. Environmental scientists have been saying for some time that

Top: President Hu Jintao of China walks on a red carpet beside a Nigerian Army officer in Abuja, Nigeria, April 26th, 2006. Nigeria gave China four oil drilling licences in exchange for a commitment to invest $4 billion in infrastructure.

Above: The reason China is anxious to sign deals in Africa and elsewhere. The thirst for oil in the rapidly growing country is outstripping supply as these motorists lining up to buy petrol at a petrol station in Dongguan, south China's Guangdong province on August 17th, 2005 testify. Closed service stations, fuel rationing and queues hours

long plagued China's southern manufacturing heartland of Guangdong during the year. The shortages piled pressure on the country's oil companies to boost supply.

"While many of the factors that have caused the oil price spike appear to be fleeting, there may be no respite from Chinese demand for the foreseeable future. The country's industrial base is gobbling up vast amounts of petrochemicals... The number of cars on mainland roads – about 20 million – is expected to increase by 2.5 million this year alone."

Matthew Forney, Time Asia,
Monday October 18th, 2004.

Above: In a throwback to the car-selling methods of the 1960s and 70s in Europe and North America, Chinese models pose with cars at the opening ceremony for the new Bentley Beijing showroom on June 1st, 2002. In Europe these selling methods would now attract derision or even demonstrations but they seem to work in China, at least temporarily. Although there is a vast and growing market for cars, and increasing traffic jams in most cities, both domestic car manufacturers and importers are desperately trying to sell more.

the global economy is being slowly undermined by the trends of environmental destruction and disruption, including shrinking forests, expanding deserts, falling water tables, eroding soils, collapsing fisheries, rising temperatures, melting ice, rising seas and increasingly destructive storms."

He said that despite these predictions a large number of economists and politicians remained to be convinced that civilisation was under threat. The development of China would change all that. For some 30 years the United States with less than 300 million people, less than 5% of the world's population, had consumed one third of the world's resources. That was now changing rapidly as China, with 1.3 billion people, had overtaken America in the number of mobile phones, television sets and refrigerators. As a nation they were already eating twice as much meat, and using twice as much steel.

China is continuing to expand consumption fast. If it reached the same living standards as the United States, as it aspires to do, then in 30 years its projected 1.45 billion people would consume the equivalent of two thirds of the 2005 world grain harvest. China's paper consumption would also be double today's world production. China only wants one car for every two people but one day if it has three cars for every four people, US style, it will have 1.1 billion cars. The whole world in 2006 has 800 million cars. To provide the roads, highways and parking lots to accommodate such a vast fleet, China would have to pave an area equal to the land it now has planted with rice. It would also need 99 million barrels of oil a day. Yet the world currently only produces 84 million barrels per day and is unlikely to be able to increase this by much.

To make his point Lester Brown was going beyond current Chinese aspirations for its people, but even if they got half as affluent as the average American their consumption demands would devastate the world economy.

Yet as he pointed out, the Chinese plan is to provide for its citizens the fossil-fuel-based, car-owning, throwaway economy that most of the developing world also aspires to. India, which by 2031 is projected to have a population even larger than China's, appears to be aiming in the same direction. So do the 3 billion other people in developing countries whose politicians promise them policies of expansion and consumption on the American pattern. The problem with this model is that all countries are competing for the same oil, grain and steel. Fortunately Lester Brown is not the only one who can see that this cannot go on. In fact it has been common knowledge since 1972 when the first United Nations Conference on Human Development was held. Professor Barry Commoner, a distinguished American biologist, told the first day of the conference: "We know that natural systems which support our life cannot long withstand the wasteful destruction of the Earth's irreplaceable stores of fuel and metal consumed by factories." It was the first time that the idea of sustainable development was discussed, the idea that the earth we leave to our children should be in at least as good a state as the one we inherited. The idea of harnessing solar power was high on the agenda. So was the argument, from America, that cutting back on use of resources would cost jobs and damage living standards. Not much seems to have changed.

Yet 34 years later Lester Brown believes that he can see the beginnings of a new-style economy.

Left: Although there is a growing gap between the cities and the poor in the countryside, western-style consumerism has come to China, in the form of boutiques and superstores. Here two streets, which could be anywhere in China's many mega-cities, are crowded with shoppers anxious to use their new-found spending power. The streets are (top) in Chengdu and (bottom) the enormous Sun Dong An Plaza on Wang Fu Jing Street in Beijing.

Following spread: While China's new urban affluence fuels a consumer boom in the rapidly growing economy, the menace of the encroaching desert is never far away. Here the legendary Forbidden City and behind it the Great Hall of the People are seen in Beijing on April 11th, 2006 through the choking atmosphere of a dust storm.

Floating dust from northern China's Inner Mongolia region frequently brings hazardous air pollution to the country's capital.

He saw hope in the wind farms of western Europe, the solar rooftops of Japan, the fast-growing hybrid car fleet of the United States, the reforested mountains of South Korea and the bicycle-friendly streets of Amsterdam. "Virtually everything we need to do to build an economy that will sustain economic progress is already being done in one or more countries," he says.

So although the problems have got a lot worse since 1972 at least some of the solutions have been developed and are available commercially. Just to take one example which Lester Brown mentions, the petrol/electric hybrid cars on the US market in 2006. The average new car sold in the United States in 2005 achieved 22 miles to the gallon, compared with 55 miles per gallon for the Toyota Prius, a dual petrol/electric car. If the United States replaced its car fleet over 10 years with efficient petrol/electric hybrids, oil use could easily be cut by half. It would be a huge start on Mr Bush's aim to reduce the US economy's dependence on oil.

What is clear, even at the current rate of consumption, is that the lifestyle of the earth's existing human population is not sustainable. Take one example, London, which draws in food and resources from across the world to feed its population of around 7.5 million. A calculation in 2005 showed that London uses the environmental resources of an area 120 times its own size to feed and service its population. That is an area greater than all the productive farmland in Britain. Around 80% of the food consumed in the capital is imported, making the prices and availability of food supplies vulnerable to a increase in oil prices.

London is not a typical city because its inhabitants include some of the richest on the planet. But half the world's population live in cities, some of them three times the size of London. All of them need vast areas of agricultural land and resources to sustain them. This realisation has led to a number of world cities getting together to try to avoid the consequences of climate change and over-consumption of the earth's resources. All have similar problems while at the same time being completely different, even when they are geographically close together. In Europe they include Barcelona, Berlin, Copenhagen, London and Paris; in Asia, Beijing and Tokyo; and in North America New York, Chicago, San Francisco and Toronto. Mexico City, probably the largest urban sprawl in the world, and Cape Town in South Africa are also involved — in the common knowledge that they cannot go on as they are. Fifteen city mayors and their teams of officials met in London in 2005 to pool knowledge and ideas on how to sustain themselves in the 21st century without destroying the planet in the process. Many of them, as has been discussed in previous chapters, are in danger from sea level rise. Others are short of water. All of them agreed they used far too much fossil fuel. This last problem dominated the conference: the need to cut greenhouse gas emissions, while at the same time transporting citizens and providing heating, cooling and light. All of them had targets to reduce emissions, but a remarkable variety of methods of doing so. Nicky Gavron, the deputy mayor of London, said the most rewarding part of the conference was the number of ideas and methods of improving quality of life that could be transplanted to other cities while at the same time cutting carbon dioxide emissions. Many schemes had proved themselves in one city but had never been thought of or tried in another. London's congestion charge, which has reduced traffic by 30%, lessened air pollution, and improved

238

Top: Even the most die-hard of American car manufacturers realise that the days of the gas-guzzling models are over. While vehicles that use five times as much fuel as European cars are still big sellers in America the market is shrinking rapidly and there are waiting lists for greener models. Here in Detroit, Michigan in 2004 Toyota unveiled its hybrid gas-electric Highlander SUV at the motor show, stealing a march on the entire US industry.

Above: The engine of the all-new Prius in Tokyo, April 17th, 2003 gets a stand all its own. The new Prius is equipped with Toyota's hybrid system THS II that allows the car to run on either petrol or electric-ity, as well as a system in which both the petrol engine and the electric motor are in operation at the same time. This is all done automatically by the car in response to the conditions and use of the brakes and accelerator by the driver.

"If economic progress is to be sustained, we need to replace the fossil-fuel-based, automobile-centred, throwaway economy with a new economic model."

Lester Brown, president and founder of the Earth Policy Institute, Washington, presentation to OECD, Paris, February 2006.

Above: Bicycles still out-number any alternative means of transport in China (apart from walking) and even in the booming city of Shanghai it remains the most popular form of get-ting to work. Try to imagine what would happen if only a small percentage of these cyclists took to cars. The entire city of 16 million people would clog up.

public transport was one of the examples. Traffic-related carbon dioxide emissions inside the congestion zone in London had been cut by 19% and bus usage in the city had risen 40% in five years. Some cities had already embarked on large scale energy efficiency programmes. New York City had already completed 164 projects with annual energy savings of $14 million (£7.5 million). Toronto in Canada had developed a deep water lake water-cooling project to air-condition large office buildings. Retrofitting 467 privately owned buildings with energy efficiency appliances and materials saved $102 million (£55 million) in energy costs in the city.

Beijing, already conscious of its vulnerability to water shortages, and anxious to improve its air pollution problems ahead of the 2008 Olympic Games, planned to cut coal burning in the city to less than 15.2 million tonnes in 2007 from 26.4 million tonnes in 2001. Solar, geothermal and wind energy are being employed to provide an 80% "clean and efficient" energy mix by 2010. But during 2005 the city had an additional 230,000 cars on its roads. The city countered this extra pollution with 2,100 buses running on compressed natural gas and before the Olympics will have 90% of its buses and 70% of its taxis running on clean fuels.

In order to try to match and exceed some of the targets of other cities London has employed an engineer called Allan Jones to run the London Climate Change Agency. In energy circles he has become something of a legend by transforming a small English borough, Woking in Surrey, into the alternative energy capital of Europe. By using energy efficiency, solar roofs, combined heat and power, and the first fuel cell power station in Europe, he reduced by 40% the borough's energy needs in 10 years and saved the council £4.7 million ($9 million) in energy

Left: Boats pass by Shanghai's skyscrapers in the Pudong financial district on October 10th, 2001. The country's showcase commercial centre has bold ambitions to become a regional financial centre. Shanghai spent a year giving itself a massive makeover in order to stake a claim as China's commercial powerhouse. But as the picture clearly shows, the city is vulnerable to climate change; the base of these vast skyscrapers are only a metre above the water level. Their basements will be inundated by sea level rise well before the end of their planned lives unless protected with new flood defences. The city is also sinking as the water table drops.

"We see the UN Framework Convention on Climate Change as the only global forum for action on climate change."

Su Wei, deputy director for treaty and law in the Chinese ministry of foreign affairs, commenting on US attempts to scupper the Montreal climate talks, December 9th, 2005.

bills in the process. The annual savings on the council's budget were £725,000 ($1.3 million) a year.

London has the target of reducing 1990 levels of carbon dioxide by 20% by 2010, in line with the national target. However, by the same time, as a demonstration, London wants to establish one zero-carbon development in every one of its boroughs. Since 70% of the city's emissions come from buildings, retrofitting existing buildings is going to be a major part of the programme. By 2010 London's alternative energy developments are expected to generate £3.35 billion (more than $6 billion) worth of business and create 7,500 jobs.

The London conference of major cities was part of a much wider network of leaders of local government who are signing up to climate change targets. An organisation called ICLEI, which represents local governments of all shapes and sizes, has signed up to meet and beat all national targets, wherever their members are. ICLEI, which calls itself local governments for sustainability, has a membership of 475 councils in dozens of countries. At the Montreal climate talks hundreds of them signed up to a target of 30% reduction in greenhouse gases by 2020 and 80% by 2050.

Also in Montreal was Greg Nickels, the mayor of Seattle. In March 2005 he persuaded eight other mayors of US cities to join him in writing to 400 colleagues across the country asking them to sign up to meet the US's Kyoto targets, even though they had been repudiated by President Bush. He was astonished by the response he received from mayors of all parties. By May 2005 he had signed up 134 mayors in 34 states, all agreeing to reduce their city's carbon dioxide emissions by 7% on 1990 levels by 2012. In Montreal the number had grown again and it passed the 200 mark in February 2006.

This kind of commitment by local government is reflected in the surge of orders for renewable energy. While oil and coal production expanded 2% annually in the first five years of the 21st century, wind and solar energy have grown by more than 30% per year. Many mayors are finding that solar energy and small wind turbines fixed directly onto the buildings that need the power are far cheaper than buying electricity from the grid.

Left: China has laid great emphasis on making the 2008 Olympics in the city "green". Coal-fired power stations are being shut down and renewables encouraged all over the city. Polluting public vehicles are being replaced and, as shown here, the organising committee for the 29th Olympics is planting trees in China's capital.

Boom Time for Jobs and Sunrise Industries

Previous spread: Denmark pioneered large scale uptake of onshore wind energy by giving people tax breaks to install their own turbines and communities incentives to invest. That success spurred the government to look at the shallow seas round Denmark's coast to see if offshore wind farms were also viable. The wind is less gusty offshore and tur-

bines can be bigger, so producing more power than previously expected. This one is in the Baltic Sea near the capital, Copenhagen. As a result of the skills engineers have developed Denmark now has a large wind energy industry, a big exporter, employing 20,000 people. The country now supplies 20% of its energy from renewable sources.

Above: The Tarbela Dam in the Indus river in Pakistan was built to provide irrigation and hydroelectricity for the country and was the largest earth-filled dam in the world when it was built in 1977. It has been very controversial and like many large dams did not live up to expectations. Its life has also been shortened by rapid silting up because

deforestation upstream has brought large amounts of sediment into the dam during the annual floods. Misuse of irrigation water has also caused salination of soil and loss of crop-lands, even desertification in some once fertile places. Loss of glaciers will severely disrupt the summer flow of the Indus.

Above: The solar panels of the world's biggest roof-based solar system are seen here in the southern German town of Bürstadt when it opened on May 24th, 2005. The 40,000 square metre installation produces 4,500,000 kilowatt-hours per year and is one of many solar ventures in Germany, which has embraced the technology alongside wind power. The government aim is to get the country's greenhouse gas emissions down to 21% below its 1990 levels by 2012 in order to meet its Kyoto target.

Developing technologies for generating electricity and replacing fossil fuels is the greatest single business opportunity of the 21st century. The race is on to refine existing and new technologies that will use the minimum amount of coal, oil and gas and eventually replace them altogether.

Despite the failure of many governments to give more than the minimum incentive to bring about a new industrial revolution it is going to happen anyway. Smaller countries like Denmark, Iceland and most recently Sweden are leading the way. The continued resistance and smoke screen thrown up by the fossil fuel lobby over the last 20 years cannot any longer prevent progress. Even George W Bush, who has done most to hold up change, acknowledged in 2006 that his country needs to be weaned off its dependence on oil.

The good news is that the technologies to change the way the world fuels transport, keeps the lights on, cooks, and keeps cool and warm already exist. All that has been lacking has been the political will to develop and modify each system to make them competitive with fossil fuels. The right incentives would promote large scale production and make renewables mainstream industries, wiping out the advantages of the still heavily subsidised fossil fuel industries. There is no magic bullet, but lots of different solutions. Research and development are still throwing out surprises. Which is the best technology often depends on where you live, what your needs are, and what natural resources you have around you.

Denmark, as is discussed shortly, has made the most of wind power, but is also experimenting with using numerous technologies to make one of its islands entirely fossil fuel free, simply to see if it is possible. Iceland wants to become the world's first hydrogen economy by 2050. This means powering all its boats and cars with hydrogen produced by electricity from surplus renewables — wind, geothermal power and hydroelectricity. More recently Sweden has announced a bold plan to wean itself entirely off oil in 15 years. For a cold country, with 9

million people, most of whom drive cars, this is a tall order. The Swedish government simply said it was to replace all fossil fuels with renewables before climate change destroys economies, and growing oil scarcity leads to huge new oil price rises. Sweden has a head start on other countries because by 2003 26% of all energy already came from renewable sources and this was rising all the time. This compares with 6% in the EU as a whole. While hydropower is the main source, Sweden also has geothermal, wind and solar power. Its main future source of local renewables is the forests. Large quantities of waste can be converted into bio-fuels.

Political leaders and commentators never stop talking about globalisation and it is a matter of astonishment how slowly, and how little, good and easily adaptable technologies travel. Some countries have been using cheap and easy renewable sources of power and energy for years while 100 miles away, across a national boundary, the same resource is ignored completely. In a global economy where cheap fruit, vegetables, televisions and textiles cross the world in ships and planes, technologies to produce renewable power or reduce greenhouse gases have seemed unable to travel.

That may be about to change for a rather surprising reason: a sophisticated and to many eyes bizarre scheme called carbon trading. This is a difficult idea to get the brain round, involving as it does people buying and selling pollution. In simple terms the scheme works because some governments set limits on how much carbon dioxide companies can produce in their manufacturing processes. If companies get more efficient, or invest in advanced processes to produce less carbon, they get credits in the form of tonnes of carbon dioxide saved. Other companies that fail to invest, or for other

Above: Forests have been one of Sweden's assets, providing wood for construction and paper-making. There has always been a great deal of waste in wood production but with modern technology all the offcuts and smaller branches and needles of these pines can be turned into bio-fuels. The Scots pine forest at Gävle, Sweden will be managed as part of the government's plan to wean the country off its dependence on oil entirely in 15 years.

reasons do not reduce their emissions, and so exceed their government's set targets, face a fine for every tonne of carbon dioxide they emit over the limit. These companies have a possible route of avoiding the fine, this is to buy tonnes of carbon from a company that has a surplus, hence the carbon trading scheme.

As part of its effort to reduce emissions under the Kyoto Protocol the European Commission introduced in 2005 a carbon trading scheme throughout the 25 countries in the European Union. To get some idea of the size of the scheme, it covered 11,428 factories and other installations, allowing a maximum of 6.6 billion tonnes of carbon dioxide to be emitted over the three years 2005-2007.

At first a tonne of carbon dioxide emissions cost as little as €5 (£3.40, $6.30) on the market but by April 2006 had reached €30 (£20, $39) a tonne. Although it subsequently fell dramatically again and prices proved volatile the €30 price tag meant the market was potentially worth over €60 billion (£40 billion, $75 billion).

But there is another, and to me rather unexpected, development of this system. When the Kyoto Protocol came into force, as has been previously mentioned, it was possible for industrialised countries to claim credits for developing clean technologies in developing countries. This was done on the basis that the atmosphere benefited wherever the carbon was saved.

European companies, some of them part of multinationals with headquarters outside Europe realised that this scheme could be combined with carbon trading to make money. Faced with expensive investments in Europe to reduce carbon dioxide emissions to meet government targets, companies have opted instead to install cheaper clean technologies in developing countries, notably China, and claim the carbon credits in Europe.

The thousands of tonnes of carbon saved each year in China can then be turned into cash in Europe. After 2007 carbon saved under the clean development mechanism may be turned into credits in the European trading scheme and traded. For example if a European company installed an energy efficiency scheme in China, which cost €10 for each tonne of carbon saved, and then was able to sell the carbon credits in Europe at €30 a tonne, it would be a highly profitable venture. Of course, if the price fell to €12.5 a tonne, and then €8.5 as it subsequently did, then the profit would go from marginal to a loss. The drops in value were caused because carbon emissions for some countries, including Germany, were far lower than anticipated and industry had no need to buy credits on the carbon market. There was therefore a rapid drop in the price per tonne, but this is expected to recover.

Despite these difficulties carbon trading has touched off a remarkable change. Following years of slow progress in spreading clean technologies by other means, there were 700 projects globally in the pipeline in June 2006 under Kyoto's clean development mechanism, many of them exploiting the European and other trading schemes.

One of the pioneers has been the international lawyer James Cameron, co-founder of Climate Change Capital, and advisor to the Alliance of Small Island States. He operates from the West End of London and has concentrated on a greenhouse gas reduction programme in China. This is simply because "it has huge opportunities". One of the first schemes was at

Left, top: Maki Uchida waters the landscaped garden of Cornell University's solar-powered house that is designed to be both functional and sustainable, in Washington, DC in October 2005. The house was designed as part of a competition between 18 collegiate teams from the US, Canada, Puerto Rico and Spain. The houses were judged in 10 areas including architecture, liveability, comfort, power generation for space heating and cooling, water heating and powering lights and appliances. Each house also had to produce enough extra power for an electric car.

Left, bottom: The University of Maryland's entry in the same competition. The house was designed to resemble the path of the sun across the sky.

the port of Shenzhen, opposite Hong Kong. Shenzhen has grown from 20,000 people to 12 million in less than 30 years, and the result of this urbanisation has been a vast refuse problem — 3,000 tonnes a day — rotting down to produce the powerful greenhouse gas methane. But for carbon trading this would have been allowed to vent into the atmosphere adding to climate change problems; but Cameron's investors paid for simple measures to capture the methane and burn it to produce energy. Since methane is 23 times as potent a greenhouse gas, tonne for tonne, as carbon dioxide the value of the investment in terms of carbon saved is very large and continues year on year. This provides a valuable income in terms of carbon credits that can be sold on the European trading system and at the same time the Chinese are getting paid to clean up their dumps.

"This is not a market for the faint-hearted because prices go down as well as up and it is a complex business setting up these deals and getting verification. But in two weeks of prospecting in China we found 100 million tonnes of potential emission reductions. It is a question of lining up economic interests with the public good," Cameron said.

Initial investment offered to his company by companies wanting to gain carbon credits by investing under the clean development mechanism was $130 million (£70 million), but he says that was "seed money. If this works for these investors there are billions waiting in the wings."

His optimism was borne out by the "remarkable growth" in the carbon market reported by Halldor Thorgeirsson, deputy executive secretary of the United Nations Framework Convention on Climate Change in May 2006.

He said it was a clear indication of the success of the Kyoto Protocol. In the five months since the Montreal conference the number of registered clean development mechanism schemes had risen from 40 to 176, and there were more than 600 in the pipeline.

Richard Kinley, acting head of the UN Climate Change Secretariat in Bonn, said: "We are presently fast approaching the one billion tonne mark in emission reductions, which corresponds to the present annual emissions of Spain and the United Kingdom combined." The two officials said that the forward-looking businesses that were prepared to invest in these carbon markets needed stability and they were hoping to convince national governments to help in this.

The positive and sophisticated approach by businesses to both comply with the regulations and use them to make money out of the problem is in sharp contrast to the weak political will to adopt winning green technologies on home turf. This is despite the fact that developed countries often have at least as much renewable potential as the countries being helped under the clean development mechanism. The United Kingdom is a classic example. Despite much rhetoric, and often repeated claims to be a world leader, the Blair government was remarkably slow to help the renewables industry, although within it Scotland and Wales are both more advanced. After three years of minimal support for renewables after coming into office in 1997 the government plumped for boosting wind power. The saga of wind power is a good example of the difference political will can make to the development of an industry, given that the UK has the largest and most reliable wind resource in Europe. Perhaps because of that incentive electricity generation using wind power was first invented in Britain. However, without

Above: Four scenes from Chongming island, which is slightly larger than Cyprus in the Mediterranean, which the Chinese government aims to turn into a showcase of sustainable development. These traditional farmers will be protected while Dongtan, a new city being built in the eastern part of Chongming island, will act as a model for the future. The planners have given promises to protect the wetland habitat of the island, fast disappearing in China. The island, in the mouth of the Yangtze River, will only have electric vehicles, while the wetlands ecosystem is protected by a special "buffer zone" separating it from the new city.

"Tides such as ours, instead of being universal, occur only in a very few places on the globe; so that if we could harness to our industries the stupendous daily rush of millions of tons of tidal water through the Pentland Firth... no Englishman need ever go underground again for fuel."

George Bernard Shaw, in his pamphlet
The Commonsense of Municipal Trading,
1908.

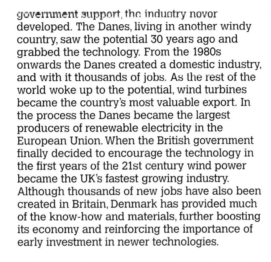

government support, the industry never developed. The Danes, living in another windy country, saw the potential 30 years ago and grabbed the technology. From the 1980s onwards the Danes created a domestic industry, and with it thousands of jobs. As the rest of the world woke up to the potential, wind turbines became the country's most valuable export. In the process the Danes became the largest producers of renewable electricity in the European Union. When the British government finally decided to encourage the technology in the first years of the 21st century wind power became the UK's fastest growing industry. Although thousands of new jobs have also been created in Britain, Denmark has provided much of the know-how and materials, further boosting its economy and reinforcing the importance of early investment in newer technologies.

At the same time as boosting wind the British government sent all the wrong signals to other home-grown renewable industries by continuing to put millions of pounds a week into nuclear power. In 2006 this was made worse by joining a new international debate about whether to build more nuclear stations (see next chapter). Despite this diversion the drive for alternative energies in the UK, and the rest of the world, is now unstoppable. This is because the UK is running out of fossil fuels and becoming a net importer of gas and oil for the first time in a generation. Even the UK's famously backward-looking Department of Trade and Industry has accepted that several renewable industries have huge potential. Young, bright civil servants have convinced at least some of the older nuclear power enthusiasts that there is room for several technologies. As a result there are already many different renewable schemes contributing to the national grid. With the right research and development

money the potential is so great that fossil fuels could be cut out entirely, although so far the government has failed to grasp the opportunity.

To give an idea of the range of renewables available to the UK is also to give a list readily available to many countries. The UK has well developed wind and hydropower generation, and a tiny but increasing contribution from solar. Geothermal, in the UK's case mainly hot water and heat from underground, is also being exploited and there is a small potential for more. Methane being gathered from landfill sites, the technology now being transferred to China under the Kyoto Protocol, was also pioneered in the UK and is the country's cheapest form of renewable power. Rubbish digesters, sewage and dung are all used to create gas to produce energy. Straw, wood waste, sewage sludge, household rubbish, chicken droppings, tyres and even condemned cattle carcases are all incinerated to generate more power. Some of these are controversial, including the burning of tyres and household rubbish, both of which many think should be recycled, but they all show that what used to be regarded as waste to be put in holes in the ground can be put to good use. But as with much else in this area the British government has failed to give clear direction or encouragement on these issues.

But with binding targets having been placed on electricity supply companies to use at least 10% renewable energy, backed with subsidies to investors, offshore wind power should continue to expand, although at the time of writing some major projects have stalled because of a rapid rise in the price of steel caused by a scarcity. The main reason for the lack of development of most of these industries seems to be the old problem of the insufficient financial incentives from government to bring new technology from

Left: This is one of only four pump storage facilities in Britain. The 440 megawatt Cruachan power station on the banks of Loch Awe in Argyll, Scotland, looks like most hydroelectric schemes producing electricity as the water stored behind the dam runs through the turbines. At slack times, however, when there is no demand for electricity, the surplus power generated is used to pump the water back up to the dam from Loch Awe. This is a way of making sure there is always plenty of water available to provide power at times of maximum electricity demand. Until scientists think of another way of storing electricity, apart from low voltage batteries, Scottish Power, the owners, believe pump storage is the most efficient method of harnessing power from renewables so that it can be used when most needed.

the successful prototype to the take-off point of mass production.

The United States is another country where the dominance of the fossil fuel industry and the government's continued subsidies for oil and coal in the form of tax breaks mean that cleaner renewable industries continue to struggle. In Britain, where the government has repeatedly said it is in favour of renewables but fails to support them, many expect to see the technologies developed elsewhere and then, as with wind, reintroduced into Britain. Among these could be wave power, and one of the newer technologies with the greatest potential, tidal turbines. Wave power has been in development for 30 years, and a bewildering variety of machines have been built, many of the promising prototypes not finally succeeding. The greatest problem has been how to deal with storms, and the unpredictable destructive forces they generate. The installation of the first commercial wave power generating station off the coast of Portugal in 2006, using British technology, could be the start of a revolution as important as the wind industry. This station relies on the movement of a series of devices floating on the surface of the sea to harness the power to drives turbines.

One of the least publicised but least problematical and most certain winners of the newer renewable generators is the use of undersea turbines. Some work in exactly the same way as wind turbines, using tidal currents instead of wind. Others use underwater sails to capture the force of the tide to provide the power to turn the turbines. There are two main advantages of these undersea technologies. Unlike wind, tides are entirely predictable, months and years in advance, so the output from the turbines is entirely reliable. The second

is that the turbine blades, or sails, need to be only one fifth of the size of those of wind machines to generate the same amount of energy. This is simply because water is far more dense. Tidal machines are now being developed along the Scandinavian and UK coastlines with a variety of consortiums interested in the potential. Placed on the sea bed, out of sight, they have none of the difficulties the wind industry has faced from people who believe that its turbines ruin the landscape.

So, as with wind and waves, Britain is peculiarly lucky to have such a large potential for another winning technology. The curious fact is that the UK's Department of Trade and Industry appears to do its best to play down this technology, officially claiming it could only provide 4% of the country's electricity. This is a ridiculously low estimate according to scientists and businessmen who have been studying undersea currents. They say that with so many offshore islands, most surrounded by tidal races entirely suitable for undersea turbines, this renewable resource could make a large contribution to Britain's energy needs well before it is possible to build nuclear stations. Among the places that have already been surveyed are the Channel Islands, which have far in excess of the resources they need for their own total electricity production. This raises the possibility of using surplus electricity for other purposes or for export. It is only a short distance, via cable, to supplying mainland France. The Bristol Channel, which has some of the highest and strongest tides in the world, could supply both south Wales and the English west country. The only missing ingredient to revolutionise Britain's power industry and make the country a net exporter of power is the political will to make it happen. Why this is so is a mystery, although the mood could change, as happened with wind.

Left and above: Vehicle congestion and the pollution from the exhausts of cars, taxis, buses and lorries are both economically damaging and the cause of ill health and premature deaths in thousands of people each year. After years of inaction local governments and city mayors are at last taking action to protect the ordinary people who live in their cities. Ken Livingstone, the London mayor, took the bold step of introducing a charge for private cars entering the city centre. The congestion charge cut traffic, reduced pollution, and allowed public transport to move more freely, encouraging far more people onto buses. Other cities in Britain are considering similar action.

Many cities now use buses running on natural gas. London buses have also switched to cleaner fuels. Air pollution has fallen dramatically as a result.

Perhaps the strangest aspect of the climate change debate over the last 20 years is that many of the easiest options for cutting fossil fuel use, which save money and reduce greenhouse gas emissions, have been around for decades yet remain seriously neglected. Everyone mentions energy efficiency as the first priority but few countries have taken it seriously. It is an example of where industries have quietly gone ahead without a political lead. British Telecom realised it could save money by cutting energy use in the mid-1980s. Once it had devoted staff and time to considering the problem the company found it far more rewarding than expected. Ahead of its time the company set a target of reducing carbon dioxide emissions by 25% by 2010 and purchasing 10% of its electricity from renewables. By 2004 combined reductions of emissions from energy use and transport were 46%. The enthusiasm with which staff became involved in the drive to switch to renewable power and save greenhouse gases improved staff morale and has encouraged the company to attempt to supply all its power needs from renewable resources. Other examples include the bank HSBC and British Petroleum, which has spent million of pounds projecting a new image of itself as Beyond Petroleum. Although this company announced record profits from record oil prices in 2006, and is still regarded with suspicion by environmentalists because of some of its projects in Alaska and the Caspian region, its television advertising alone has done much to raise the profile of climate change in the UK. This change in a company that until the mid-1990s was a major contributor to the Climate Change Coalition, the pressure group campaigning against attempts to reduce coal and oil use, shows what can be achieved. Since 2001 the reduction in emissions from the company's energy efficiency projects has

reduced emissions by 4 million tonnes, the quantity of pollution generated by a European city of half a million people.

Subsequently the company has decided to invest £200 million ($370 million) in energy efficiency measures and expects to get more than that back from energy savings. The company already has a solar business, and as part of its "beyond petroleum" image the company has pledged to extend its global wind power capacity to 3,000 megawatts by 2015. The company says this is enough to power 1.5 million typical homes in Europe.

Compared with Exxon Mobil, the world's largest oil company, BP has made great strides in the right direction. There can be no doubt that any or all of these oil companies could invest vast sums in renewables and hardly notice it. Exxon surpassed BP Amoco and Shell by declaring the largest profit ever recorded in the world in 2005. BP and its rival Shell have made strategic decisions to invest in the future while Exxon Mobil has made large energy efficiency savings but turned its back on anything but fossil fuels and cancelled its renewables programme. BP has decided to concentrate much more heavily on gas, a far cleaner fossil fuel than oil and coal, but also makes it clear that gas is only an interim measure towards a larger renewables portfolio. There is no doubt that the personal conviction of Lord Browne, the group chief executive, at the helm of BP Amoco drove the climate change programme. It can only have helped his cause with the rest of the board that many analysts believe that the era of cheap oil is over for ever and that dwindling supplies mean that the tipping point where demand exceeds supply is close. This alone would make sense of the company's decision to invest heavily in solar power. BP is now a major player

Right: Advertising is a powerful medium. The European Commission believes that by bypassing governments which are not doing enough to combat climate change, and appealing direct to the public, it can change behaviour. It is amazing how much carbon dioxide would be saved if everyone did what the poster (top) suggests. The

Chevron poster (bottom) directs the public to its website and shows that for the first time an oil company is prepared to admit that one thing is clear: the era of easy oil is over. The company says that many of the world's large oil fields are "maturing", in other words coming to the end of their maximum production, and that new finds are harder to

exploit. The company is involved in a continuing debate with the public on its specially set up website about what to do next, a facility which the company says has been a great success.

YOU CONTROL CLIMATE CHANGE.

TURN DOWN. SWITCH OFF. RECYCLE. WALK. CHANGE

In the next 20 years, the world will grow by nearly one and a half billion people.

So how do we feed their appetite for energy?

will you join us.com

Chevron

Human energy™

"Wall Street is waking up to climate change risks and opportunities. Considerably more of the world's largest corporations are getting a handle on what climate change means for their business and what they need to do to capture opportunities and mitigate risks. This all points to a continued elevation of climate change as a critical shareholder value issue for investors."

James Cameron, chairman of the Carbon Disclosure Project, 2006.

Above: One of the hopes of staving off the worst effect of global warming while we find alternatives to fossil fuels is carbon sequestration. This giant apparatus in the Algerian desert is designed to inject one million tonnes of captured carbon dioxide gas into natural gas wells. The natural gas will be pumped out and the carbon dioxide will replace it and the well will eventually be sealed. It is hoped the carbon dioxide will be stored for hundreds, possibly thousands of years, while mankind attempts to get the atmosphere back into balance. BP is among those investing in this technology, which would help the company avoid carbon taxes, and notch up carbon credits to trade internationally.

in a market growing as fast as it and other larger manufacturers can make and deliver the solar panels. For now, though, the company, in its duty to shareholders, will continue to defend its interests in oil as long as fossil fuel remains the bulk of its business and the mainstay of the company's profits.

But another positive sign is that one of the schemes the company is pioneering is carbon storage. This involves capturing the exhaust gases from burning fossil fuels, typically from coal-fired power stations, and pumping them under pressure down old oil wells. The idea is to store carbon dioxide permanently underground, like the coal it came from. One of the target areas for this is older oil wells which are near exhaustion. As a bonus for the company the introduction of carbon dioxide under pressure pushes more oil out of the ground.

Energy efficiency is something everyone knows will work, but most do little about. The performance of these multinational companies shows that whatever business you are in, and irrespective of where you operate in the world, energy efficiency can save hundreds of millions of pounds through simple measures. The other battle is to make the best use of energy produced by fossil fuels. This involves the widespread adoption of combined heat and power. That is a simple system where the surplus heat from generating electricity is used for district heating rather than allowing it to be wasted by letting it escape into the atmosphere. Combined heat and power plants still work well on a large scale in former eastern bloc countries. They are even more efficient with small generators in factories, groups of office buildings or housing estates elsewhere. In some so-called advanced countries, notably Britain and America, this is a badly under-utilised

technology. It is among the many examples across the world where some straightforward methods of saving fuel and carbon emissions have been embraced in one country and ignored in others, when they could be universally effective.

One of the countries which has made most effort over a long period to be more energy efficient is Japan. The government began looking at alternatives to fossil fuels half a century ago. This was driven by Japan's vulnerability to oil shortages; it has no reserves of its own. Initially government and industry concentrated on energy efficiency — such devices as low energy light bulbs and combined heat and power. Ever since the 1973 oil shock Japan has been the most energy efficient large country in the industrialised world. In the last 10 years, in a bid to meet its demanding 6% Kyoto reduction targets for greenhouse gas emissions, the country has made another series of new strides.

Despite its energy efficiency reputation Japan, along with the rest of the world, still acknowledges that it wastes far too much energy because of inefficient lighting, heating and cooling. With the glass tower blocks that dominate central Tokyo cooling is now far more important than heating. Since the Kyoto agreement was signed the largest combined heating and cooling system in Japan has begun operating at Harumi Island, in Tokyo. Triton Square is a building complex housing thousands of office workers facing Tokyo Bay. The system is based on a water tank the size of 50 swimming pools. Heat exchangers provide hot and cold water as required using the massive capacity of the tank to store unused energy in the water. Electricity is only used at night, when tariffs are lower, to top up the system if necessary — further cutting energy

Above: Lord Browne, chief executive of British Petroleum (BP), has done more than any other oil executive to acknowledge the dangers of climate change and insist that his company do something about it. His re-branding of the company as "beyond petroleum" with this new sunburst corporate logo, was followed by a massive advertising campaign. He said the helios mark, named after the sun god of ancient Greece, showed a commitment to the environment and to solar power.

Above: The remarkable magnetic levitation train developed by Central Japan Railway Company. The train is held above the track by the opposite forces of giant magnets and also propelled along by a linear magnetic motor. It is capable of remarkable speeds because there is no friction but despite an 18.4 kilometre test track in Tsuru west of Tokyo, $2 billion investment and 40 years of development the railway company has yet to persuade the government to build a commercial line between Tokyo and the western city of Osaka.

costs. It has cut carbon dioxide emissions by 60% and increased energy efficiency by 40%.

While Europe has led on wind power, the Japanese have been concentrating on solar. The huge Sharp company has long been the world's largest producer, installing its first solar powered lighthouse in Nagasaki in 1963; now there are 589. Solar electricity costs have tumbled to one third of their price in 1993 and are still dropping as the technology improves and mass production reduces unit price. Exports to Germany are keeping Sharp's factories at maximum production although Germany has now developed an industry of its own.

Japan, along with other countries with volcanic regions, has also long exploited geothermal power; it is not a new technology, Italy pioneered producing electricity from geothermal power in 1905. Hot rocks near the surface, stable enough to allow water pipes to be installed, are exploited to produce steam to drive turbines. The same technology is also used to produce hot water for district heating schemes. In Iceland 93% of homes are heated with geothermal hot water and surprisingly France, more famous for its nuclear power stations, has 70 geothermal plants producing both central heating and hot water for 200,000 homes. The same techniques can be used to heat greenhouses to grow food in the winter, and for fish farms for species that grow faster in warmer water.

But while the idea of geothermal technology is simple, it requires expertise to exploit, and much of the heat that could be harnessed is wasted, partly because it seems so plentiful and costs nothing. Realising that far more energy could be utilised the Japanese have developed turbines that can re-use steam at lower temperature and

pressure. This produces far more power with the same quantity of heat. It is estimated that Japan's geothermal potential is 69,000 megawatts, enough to produce one third of its electricity needs. Japan uses the water for thousands of spas, swimming pools and to heat hotels but it is still a badly under-utilised resource.

The United States with 2,000 megawatts and the Philippines with 1,900 megawatts lead the world in geothermal energy for electricity at the moment — altogether 25 countries use it and the industry is expanding all the time. All round the Pacific rim and in many countries where hot rocks are close to the surface there is great potential. Among the countries with an obvious resource are Chile, Peru, Ecuador, Colombia, all the central American countries, Canada, Russia, China, South Korea, Indonesia, Australia and New Zealand. A United Nations report has also found potential in the Rift Valley area of Africa, and is building a pilot plant. There are also unexploited hot rocks in the eastern Mediterranean. Without know-how geothermal is often a difficult technology to exploit, so there is further scope here under the Kyoto clean development mechanism coupled with carbon trading schemes to transfer this technology to less developed nations and save large quantities of carbon dioxide. Just to give one example, Indonesia has more than 222 million people and 500 volcanoes — more than enough hot rocks to provide hot water and power for the whole country. A grand plan to build eleven geothermal power stations exists on paper but the country lacks the resources to exploit this use of carbon-free technology.

As with the potential use of wave and undersea turbines in Europe, Japan is also looking to the oceans. It is in the seas of the tropical regions that the other great promise for alternative

Top: Japan has created and maintained a steady lead in the solar revolution, with the prime minister leading by example. In this picture taken in April 2005 the Japanese prime minister, Junichiro Koizumi, centre, makes toasts with his predecessors Yasuhiro Nakasone, left, and Kiichi Miyazawa during a luncheon held to unveil his mansion. After

three years of renovation work, the mansion features solar-power panels that cover the roof of the 7,000 square metres (8,400 square yards) four-storey building.

Above: The solar calculator is to be followed by the solar mobile telephone. It is a prototype displayed in the Wireless Japan 2005 exhibition in Tokyo and will soon be charging up without needing to be plugged in. It is one of a number of solar phones that are expected to be on the market soon.

technology currently lies. Japan has one technology, "ocean thermal energy conversion", which also has enormous potential, and not just to produce electricity. It can also produce fresh water, hydrogen and lithium, a valuable industrial material.

Essentially the idea is simple. Warm water on the ocean surface is hot enough to turn liquid ammonia into a gas, powering turbines as it expands. Cool water pumped from the deep ocean is then used to turn the gas back into liquid ammonia. Using a series of heat exchangers and pumps the system can also be used to produce drinking water and extract hydrogen from sea water. The deep ocean water contains lithium and other useful substances that can be extracted, and is also high in nutrients, something that has implications for fisheries, and the potential development of large scale ocean fish farms. The Japanese Fisheries Agency is cooperating with the Saga University, developers of the system, to see if the discharge water from the energy conversion unit can be used to stimulate new fisheries.

The power available from this source is huge, especially in the tropics where water temperatures are highest, and particularly in small island Pacific states and the Caribbean where the cost of fossil fuel imports is large, and fisheries are important. Hydrogen to power fuel cells is seen by many as the front runner to replace fossil fuels in transport. For many poor coastal countries and islands this could be the answer. The only "waste" from burning hydrogen is pure water.

Meanwhile Japan has been working on the most difficult area of transport too. As we have already described, the Toyota Prius is already well known internationally as a hybrid car that uses its braking system to generate electric power. The new improved model hardly uses petroleum at all in some driving conditions and claims a 43% reduction on the carbon dioxide emissions even against the previous model. The Japanese have also harnessed this technology to help drive their trains. Tokyo Railway's newest model of train consumes 40% less energy as a result; quite a saving since the company carries 2.7 million people a day.

Transport, with the need for alternative fuels to power vehicles, ships, and aircraft, remains one of the most difficult areas. The surge in oil prices has begun to concentrate minds, not just at government and industry level, but also among consumers. The financial troubles of some of the world's great car makers like Ford and General Motors are as much due to building the wrong kind of gas-guzzling models as to the high wages and pension schemes of the workforce, which managements highlight. As with electricity production, the race has been on to find alternatives to petrol and diesel. As often happens in a world dominated by big business the most promising interim solution comes from the developing world, in this case Brazil.

Brazil is the world's largest producer of ethanol, an industrial alcohol based on waste products from the sugar industry. Now more than half the cars sold in the country can use it as fuel. The country burns 4 billion gallons annually. Four in ten vehicles run entirely on ethanol and the remainder use blends of up to 24% ethanol and 76% petrol. The country has taken 30 years to develop the technology, which began as a result of the 1970s price rise for oil. Exports of both the fuel, and the plant to make it, are a potentially big money earner for Brazil as well as providing a renewable method of running vehicles and avoiding oil imports.

Left: China has relocated about 1.3 million people from the areas along the Yangtze river affected by the mammoth Three Gorges dam project intended to tame the river and generate vast quantities of electricity. This is probably the most controversial big dam project of all time, with many doubting that it will ever live up to Chinese government expectations, partly because of the river's silt problems. The loss of so much land and the wildlife of the river is also lamented. Chinese officials say it is an alternative to building many coal-fired power stations, which would otherwise make climate change far worse. The dam came into operation on 06/06/06.

"The stone age did not end for lack of stone, and the oil age will end long before the world runs out of oil."

Sheik Ahmed Zaki Yamani, the Saudi oil minister in the 1970s.

A variety of fuels produced from plants, trees, wheat, maize, sugar cane and oil seed rape producing everything from alcohol to cooking oil is now being used to power vehicles. It is perhaps ironic that in 1898 when Rudolf Diesel first demonstrated his compressor ignition engine at the World's Exhibition in Paris he used peanut oil to power it. The oil was the world's first bio-diesel. Vegetable oils were used in diesel engines until the 1920s when an alteration was made to the engine to enable it to use a residue of petroleum, and bio-diesel was all but forgotten.

Ethanol, which is principally alcohol produced from sugarcane, can also come from a variety of grains including wheat and maize. Bio-fuels are being introduced across the world from a variety of local crops. In Europe and America they are mostly a 5% to 10% top-up to existing diesel or petrol but in many countries some vehicles run entirely on non-fossil fuels. European legislation has driven a market for bio-fuel by insisting that they take up 5.57% of all transport fuels by 2010. But Brazil being so far ahead of the game is beginning to cash in on its technological lead with an export business to China, which wants to reduce its dependence on oil too.

One of the big problems of this bio-fuel revolution is that many of the crops that are used are also food crops. This has the potentially disastrous incentive to cut down even more forests in order to provide land to grow both bio-fuels and more food. A better alternative is to find crops that can be turned into bio-fuels that are not food crops and grow on land that would otherwise not be productive. Inedible vegetable oils, mostly produced by seed-bearing trees and shrubs, provide an alternative. One example is jatropha curcas, a tree that grows in tropical and sub-tropical climates across the developing world. It happily grows on non-agricultural and marginal land and produces seeds which can be crushed to produce vegetable oil. It can be used instead of mineral diesel or mixed with it, producing a much cleaner-burning fuel. This is a tree that already grows widely in Africa, India, South East Asia and China, often in arid lands, and requires little water or nutrients. The tree matures after only three years and lives for 30, providing potential for early returns. One company, D1 Oils, floated on the Alternative Investment Market of the London Stock Exchange in 2004, has already gone into production.

The advantage for countries, regions and individual farms that invest in bio-fuels is that equipment for producing bio-diesel is both small scale and portable. It is therefore possible to hire a bio-diesel plant to arrive to coincide with your harvest, to turn crops into a cash product on the spot. The technology could make large scale oil refineries redundant.

This typifies one of the most significant battles, both in electricity generation and other technologies. It is the tussle between large centrally produced power and local solutions. Governments and companies have always been in love with the grandiose. Giant hydroelectric dams and vast concrete nuclear stations go along with motorways, bridges and sports stadiums as proof of political potency. The new generation of technologies, which offer the best hope for the future, are the opposite. Some of these are discussed in the final chapter, and involve some of the options for individuals who want to save the planet. This does not mean that big business and large scale manufacture are not needed. The idea is to produce electricity as close as possible to the point where it is used.

Left: Workers load palm oil fruits on to a lorry at a plantation near Kuala Lumpur in Malaysia. Prices for Malaysian crude palm oil continued to rise in 2006 as they were used for the production of bio-diesel, one of the world's fastest growing industries. Although bio-diesel cuts the need for fossil fuels and reduces carbon dioxide emissions it can have a down side as more tropical rainforests are felled to create new palm oil plantations.

Above: In Brazil it is the sugar cane that is converted into another alternative to fossil fuels, ethanol, in which the country has a world lead. It is one of the cheapest and most dependable fuels derived from renewable sources. Three-quarters of the cars now being produced in Brazil can run on either ethanol or petrol, or any mixture of the two. The top picture is of a plant producing new fuel, and below it are some workers taking a break after harvesting a field of sugar cane at the Sao Tome plantation in the southern Brazilian state of Parana.

At present, in most of the world, power plants are so far away from the customer that 10% of the power, and sometimes as much as 15%, is lost in transmission and all the heat generated is vented to the atmosphere. One of the major advantages of solar panels, fuel cells, small scale hydropower, mini-wind turbines and combined heat and power plants for individual homes is that mass production brings the unit price down for each customer to install it. The second advantage of these systems over the large scale is that the distance between the generation of power and point of use is metres not miles. Solar panels and combined heat and power plants do not need a grid, and all the power produced is used either directly or by near neighbours. In all grid systems, however good the cabling, some power is lost. The other possibility with renewables, which by definition do not rely on fossil fuels, is that the raw material — energy from the sun, wind, waves or heat from the ground or oceans — is free. That often means that even when the power is not directly needed for electricity to keep the lights on, or for heating, power is still being produced. Since electricity cannot be stored, except in batteries, this could be a disadvantage. It also raises all sorts of possibilities, and potential advantages, although application requires ingenuity. Perhaps the simplest example of use for spare power is for family cars. Electricity can be used directly for an electric car, and there are more and more of them on city streets. Many believe the biggest potential is to produce hydrogen by electrolysis at night from wind farms when power is not needed or in summer when there is excess solar power. Obviously producing hydrogen in this way is an expensive and not very sensible use of electric power if it is produced from fossil fuels. If, on the other hand, hydrogen was automatically produced when electricity from wind power, solar or wave

power was surplus to requirements then hydrogen would be able to replace fossil fuels in any number of applications. Hydrogen fuel cells for cars are in development. Fuel cells can also be used to produce electricity. With their by-product being pure water fuel cells would seem to be an ideal technology for desert kingdoms, currently reliant on oil. Solar panels could produce electricity, with the surplus power used for the production of hydrogen, which would power the turbines at night producing water as a by-product.

There are many possibilities of harnessing these readily available technologies but they require engineering skills to understand the possibilities. But to bring the unit price down so that individuals and local communities could take advantage, politicians and companies need to combine to ensure large scale mass production.

Above: On many remote Pacific islands electricity has not been available unless islanders buy expensive and noisy generators. Then they have to import and burn regular supplies of fossil fuels which may be beyond their means. They can, however, leapfrog the technology of the 20th century with photovoltaic panels which turn the plenti-ful sunshine into electricity without breaking into the natural sounds of the waves breaking onto the beach. This solar panel on Tobi Island, Belau Islands, in the Pacific is one of several that produce all the electricity used on the island.

Right: Solar cooking is catching on in many parts of the world, particularly in the tropics where the sun is strong. Here students in Brazil test the heat given off by one of the demonstration stoves provided by a Greenpeace Positive Energy Tour in 2004. The basic idea is to use shiny or polished steel as a mirror to concen-trate the sun's rays onto a cooking pot or kettle. The pot boils almost as quickly as over a fire and it is possi-ble to cook a complete meal without using any fuel, or other energy source, apart from direct sunlight. It also saves families tramping miles for fuel wood.

Voodoo Economics

Previous spread: Bulgarian schoolchildren put on their gas-masks during an exercise in the town of Kozloduy near Bulgaria's nuclear power plant, some 200 km (125 miles) north of the capital Sofia on October 9th, 2002. What use they would be in the event of an accident at the Soviet-made nuclear power plant is not clear, but the safety of the plant is a cause for concern. The future of the plant, which accounts for up to half of the country's electricity output, is a sticking point in talks with the European Union. The EU would like the plant closed while some Bulgarian politicians are making it a matter of national pride to insist on building a second one.

Above: Perhaps the most controversial nuclear plant in Europe, the sprawling Sellafield site on the Cumbrian coast in England seen from the local village graveyard at Seascale. The site is the home of two nuclear reprocessing plants, which produce plutonium and depleted uranium from spent nuclear fuel. Although a very small quantity is re- used for a nuclear fuel called MOX (mixed oxides of plutonium and uranium) the rest is stored on site under armed guard. The site was the scene of a nuclear fire in 1957, which left behind a contaminated nuclear reactor that is still too dangerous to demolish. Large quantities of nuclear waste are stored on site.

Above: France has the greatest reliance of any country on nuclear power and uses its major rivers for cooling water for some of the many stations. Bugey, on the banks of the Rhône river in Saint-Vulbas, near Lyon, is one of many inland that are vulnerable to being shut down during droughts and warm weather. The plant warms the Rhône so much that life in the river is threatened. During the 2003 heat wave state-owned Electricité de France cut its power capacity by 4,000 MW as a result. This was despite the fact that the government had previously relaxed environmental rules allowing plants to put the water they use back into rivers at higher temperatures than usually permitted.

Building large numbers of new nuclear power plants is seen by many as the solution to global warming. Indeed the industry has long pushed the argument itself. It is, they say, the only large scale power-producing technology that does not emit carbon dioxide.

This is one of the most contentious issues surrounding climate change. Many countries have rejected the technology entirely, while at the other extreme France has built so many nuclear stations that 79% of its electricity comes from that single source. The problems of what to do with the waste, the potential for developing nuclear weapons from the same technology (which has caused so much trouble with Iran) and the vulnerability of stations to terrorism all add to the sharp debate.

Some scientists, including the UK's chief scientific adviser, Sir David King, have urged returning to nuclear power as a solution. He believes that the UK cannot meet its carbon dioxide reduction targets without new nuclear build. Partly as a result Tony Blair, the prime minister, ordered a new energy review in Britain in 2006. Political commentators claim the only object was to overturn a study less than four years old which concluded that nuclear power was not the answer. The new study duly decided that a new generation of nuclear stations was needed, but only time will tell whether this conclusion will survive much beyond Blair's premiership.

The industry itself, on both sides of the Atlantic, has long been in the doldrums. Since the Three Mile Island incident in America in 1979 and the catastrophic Chernobyl disaster in the Ukraine in 1986 there has been public fear of nuclear power. Apart from voter opposition the world's big financial institutions have been unwilling to back new nuclear build. Without massive state subsidies for disposing of waste and insurance, the industry cannot compete. Professor James Lovelock, the inventor of the Gaia theory, thinks that the danger from man-made climate change is so great that new nuclear build to solve the problem is the lesser of two evils. The main

environmental argument against this is that a station may last 40 years but the radioactive waste produced remains dangerous for many thousands. So far, despite 50 years of trying, no solution has been found to the problem of how to store or dispose of the waste safely, with the possible exception of Scandinavian countries. They use purpose-built deep rock caverns. Everywhere else it is stored in a variety of concrete bunkers, all of them with a much shorter life than the waste. This waste will remain dangerous long after the concrete has disintegrated. There is a strong chance that climatic conditions will have changed so much within 300 years that the remaining radio-activity will be released, with unknown effects on future generations.

But the argument that will kill large scale nuclear, or should, is that it is not economic. There are lots of other better and cheaper solutions. That does not mean that the economic argument is an easy one to make. The nuclear industry still suffers from what its detractors call "voodoo economics". The sums put forward by the industry certainly do not tell the whole story, and do not add up. What is astonishing is that spurious figures are frequently swallowed hook, line and sinker by governments and journalists alike. At present the industry claims it is almost competitive with coal, gas and wind power, and if enough stations were built, economies of scale would bring the price down further. Better still, if the fossil fuel generators were made to pay a carbon tax to make up for the damage they do to the environment nuclear would be even more competitive. As fossil fuels get more expensive, as they must, the industry says, nuclear becomes even more attractive. The industry also points to the recent row between Russia and Ukraine on the price of gas, which caused an interruption to the supply pipes to Europe. Nuclear would be a

Above: A view of the dormant cooling towers of the Unit Two reactor of the Three Mile Island nuclear power plant from the visitor centre, near Middletown, Pennsylvania, in 1999, the 20th anniversary of the worst US commercial nuclear accident. The accident did not release radioactivity into the environment but members of the public are still trying to get an official admission that what happened at Three Mile Island was a meltdown. Whatever happened, it effectively stopped the United States nuclear industry in its tracks. President George W Bush has tried to revive nuclear power but there is still little enthusiasm outside the nuclear industry.

"Nuclear power is too expensive and too dangerous. Wasting billions on more nuclear reactors would distract from the real task of developing renewables and reducing energy demand. Nuclear power is the ultimate unsustainable form of energy – for a little energy now, we would be condemning 10,000 generations to deal with the radioactive waste we would produce."

Andrew Lee, director of campaigns for WWF UK, March 2006.

Above: A general view of the Chernobyl nuclear power plant in April 2006, on the 20th anniversary of the nuclear disaster there. It was the world's worst nuclear accident, contaminating thousands of square miles of Europe. Although few people were killed immediately, many are still dying as a result of the radiation released.

Top and right: Official denials of the large number of children suffering severe illnesses in the wake of the nuclear accident at Chernobyl in 1986 have made people more cynical about nuclear power, not less. Cuba has treated 18,000 Ukrainian children since 1990 free of charge for loss of hair, skin disorders, cancer, leukaemia and other illnesses attributed to the radioactivity unleashed by the meltdown years before they were born. These children do not appear in official statistics. Bagdan, a Ukrainian boy, is surrounded by friends in hospital at Tarara, outside Havana, in 2005. Natalia, also Ukrainian, receives treatment for alopecia in the same hospital.

Above: A view of an amusement park in the centre of the abandoned town of Pripyat, in the 30 km (19 miles) exclusion zone around the closed Chernobyl nuclear power plant, March 31st, 2006. About 50,000 Pripyat residents were evacuated after the disaster, taking only a few belongings.

better bet, the industry says. This ignores the fact that uranium, while plentiful at the moment, could also become expensive and more difficult to find, like oil and gas. While attacking fossil fuels, the industry correctly sees renewables as its main competitor for subsidies and capital and so pours scorn on newer technologies as being only minor players. They say that renewables will never be able to produce enough electricity to make up for the giant capacity needed, which only new nuclear build can provide. Wind turbines are particularly lampooned as being useless because the wind only blows some of the time. The industry then quotes the price of electricity produced by nuclear stations in pence per unit, making the price look competitive with all other sources of power. The problem is that this price often only reflects part of the cost. In most cases it is the cost of mining the uranium, enriching it to make fuel, and the cost of running the stations while they are producing electricity. Often omitted is the initial capital cost of building the station, which is written off, since this was often provided directly or indirectly by state aid.

Another large cost factor is the disposing of the spent fuel and radioactive waste after the uranium has been burnt in the reactor. Ironically in the past this has actually been counted as an asset since it could be reprocessed to recover spent uranium and plutonium, potentially another fuel. In fact there are thousands of tonnes of this recovered material stockpiled which has never been used. This is a dangerous liability in this age of terrorism. Dealing with this material, which the industry calls "back-end costs", is either completely ignored or heavily discounted when calculating the cost of nuclear power. This is because they will not have to be paid for many years.

In reality the cost of disposing of nuclear waste nearly always falls on the taxpayer. In the UK the part of the nuclear industry that was privatised would have gone bankrupt in 2004 if the government had not agreed to pay all nuclear waste costs until 2086, at the rate of £200 million ($370 million) a year. This fact is hidden from the public. The bill for disposing of Britain's nuclear waste was estimated in 2006 at £70 billion ($126 billion) but is rising every year. One other point that is largely ignored is that building new nuclear stations takes many years, usually more than a decade, from planning to producing the first electricity. In that time billions of pounds in capital is tied up as thousands of tonnes of concrete are poured and steel manufactured. This produces large quantities of carbon dioxide. Conservative estimates suggest that world-wide 1,000 nuclear stations would be needed to make a dent in the carbon dioxide problem. This raises the question of the capacity of the industry to build this many stations, or whether there is enough available uranium to go round. In the first round of nuclear station building, when sophisticated terrorism and suicide bombers were the stuff of nightmares and not part of the horror of modern life, there was a thriving export industry in nuclear stations. Although Russia, America, France and the UK all competed to export stations there would be many parts of the world now where providing nuclear know-how would be regarded as a security risk. Compare that with exporting renewable technology. There are no similar risks in solar, wind, hydro and biomass.

But let us examine the funding alternatives for nuclear. Take the capital cost for one nuclear station, say conservatively £2 billion ($3.7 billion) over 10 years before one watt of electricity is produced. For this sum hundreds of small scale renewable alternatives could be

Above: There are two models of the Japanese city of Hiroshima in the city's Peace Memorial Museum, which commemorates those who died when the atomic bomb was dropped in August 1945 by the US Air Force. One model is of the city before the bombing, while the second model is of how it looked afterwards. It was the first place in the world to experience the devastation caused by a nuclear explosion. The museum presents the possessions of those during the time of the bombing. The museum shows toys, books, and magazines that belonged to children in the 1940s. It is fear of this devastation happening again that makes some oppose nuclear technology.

Right, top: A warehouse containing some of the most dangerous wastes known to man, the remains of spent fuel rods from a nuclear reactor, which can remain lethal to humans for 250,000 years. Every country which has adopted nuclear power is faced with the problem of what to do with the waste. Nearly all have failed to find public agreement on somewhere to dispose of it or store it long term. This is the temporary nuclear waste storage facility in the north German village of Gorleben near Dannenberg constructed for 420 containers of waste. The German government is hoping to find a permanent resting place for the waste in old salt mines, which are said to be very stable geological formations.

Right, bottom: A worker walks past the yet to be filled nuclear waste containers, which will be stored in the warehouse. When the containers are full it will be too dangerous for anyone not wearing protective clothing to be this close to the waste.

installed, some of them within months. Solar panels, small scale hydro, and wind turbines on homes, offices and factories. They would give instant return in terms of electricity and capital employed to the places where it was invested. The renewable options also avoid the problem that nuclear stations are normally located remote from centres of population for safety reasons. This means that about 10% of the nuclear power generated is lost in long transmission lines before it reaches the customer. This is another cost the nuclear industry fails to take into account or at least to admit to the public. The industry, faced with those arguments, says that countries can build both renewables and nuclear. This ignores the point that even in the richest countries, there is not enough capital available for everything. To cut out the transmission problem, the Chinese are working on a small scale reactor the size of a dustbin. This would provide local supplies for a village or small town. They also claim it is safe because it has an automatic shutdown mechanism if it overheats.

Angela Merkel, as her first policy act as the new chancellor of Germany, decided to spend her money in a different way. Over a 20-year period all older German housing will be brought up to modern standards at the rate of 5% a year, until every home in the country is both warm and energy efficient. At a stroke she has created employment for thousands of people in the former East Germany where jobs were in short supply, and cut out the need for new nuclear build. The old stations will become redundant. The United States and Britain could adopt similar policies but politicians in both countries seem to prefer building new power generation plants. They show a preference for the giant and wasteful technologies of the last century. It always seems easier to go for the grand design

of some future plan than adopt immediate small and practical solutions. An example of doomed but currently celebrated grand design is Dubai, the small desert kingdom on the Persian Gulf. An astonishing 250,000 men, mostly from India and Pakistan, are building a forest of luxury hotels, roads, airports, artificial islands, and golf courses, believed to have a price tag of $100 billion (£55 billion). It will be the business centre and the playground in the Middle East and is all funded and fuelled by oil. Dubai used to play golf on sand, and there is still the celebrated old course, where golfers can play in traditional style, without a blade of grass in sight. But it is surrounded by lush green new courses, with attendant country clubs, as if the desert no longer existed. In one sense it does not. Desalination plants, using cheap oil from under the desert, provide the water for this lavish lifestyle. They use waste heat, and surplus power, from the electricity plants to produce fresh water. In the driest part of the world, with sometimes less than two millimetres of rain a year is the highest concentration of desalination plants in the world. More than 70% of the water in the United Arab Emirates, of which Dubai is part, comes from desalination — more than 24 million litres a day. The country has the highest consumption of water per person in the world, with the exception of the United States. Much of the water is recycled, and sophisticated sewage works mean that water is not wasted, but this new "civilisation" in the desert will almost certainly be one of the shortest lived in human history. It might be possible to build a nuclear power plant or use solar energy to replace some of the oil when it becomes more expensive, but for how long? Sea level rise will in any case overwhelm the low level artificial island paradises, so recently constructed, on which much of Dubai's new wealth depends.

Right: Germans believe that they can phase out nuclear power by adapting energy efficiency measures and building giant wind turbines like this one at Brunsbüttel near Hamburg. At the time this picture was taken in October 2004, this was the largest wind turbine in the world at 5 megawatts. In 20 years wind turbines have grown in size 10-fold, the earliest models only producing half a megawatt of electricity. This crane is lifting the blade hub 120 metres to the top of the mast. The span of the blade is 126 metres. This is a prototype being built onshore but RE Power Systems hope to install large numbers in offshore wind farms in German waters and abroad.

What Can We Do?

Previous spread: As sand storms continued to hit Beijing, and other parts of the country repeatedly suffered flooding from deforestation the Chinese government reacted by ordering what is believed to be the largest tree planting programme ever attempted. Here soldiers show the way by preparing the ground to plant trees at a Beijing park in 2001. Chinese authorities increased Beijing's green land coverage to 40% by 2005 and expect to reach 45% by 2010.

Above: These spectacular "blue marble" images are the most detailed true-colour images to date attempting to take in all of Earth from space. Using a collection of satellite-based observations, scientists and visualisers stitched together months of observations of the land surface, oceans, sea ice, and clouds into a seamless, mosaic of every square kilometre of the planet. This image shows the Americas but with a tilt towards the North Pole so that the extent of the sea ice and the huge mass of the Greenland ice cap is clearly visible. It also gives a feeling of how vulnerable and small the planet really is.

Above: The colours on these globes denote the type of landscape, or the depth of the sea. It is startling how much of the land surface is desert or a type of prairie or grassland, and how little is green. It is also interesting to note how small Europe appears to be in relation to other continents and the full extent of the Indian Ocean. Much of the information contained in these images came from a single remote-sensing device, NASA's Moderate Resolution Imaging Spectroradiometer, or MODIS. Flying over 700 km (450 miles) above the Earth onboard the Terra satellite, MODIS observes a variety of oceanic, atmospheric and terrestrial features.

It is not surprising that individuals sometimes feel powerless when faced with really big issues like man-made climate change. What can one person do when the rest of the world seems intent on self-destruction?

Politicians do not help much with this. While they exhort individuals to take responsibility for their own actions, they do little to help by changing the regulations or tax system. Still worse, one of their constant excuses when asked to do something to show leadership on climate change is that one country cannot stand alone. George W Bush took it one step further. His reason for repudiating the Kyoto Protocol was that America, and the rest of the industrialised world, would be disadvantaged against the developing world if they took action to curb carbon dioxide emissions, while others were allowed to pollute as they liked. Since the US is the world's largest contributor to greenhouse gas emissions this was abdication of responsibility on a breathtaking scale.

For individuals the excuse used by politicians for not taking action is multiplied thousands of times. What difference can one person make when neighbours and most of the rest of the world carry on burning fossil fuels at an alarming rate? Of course the honest answer to that question is that if everyone else did nothing, a personal effort would be futile. Stopping the global warming juggernaut would be impossible, even if having tried might make you feel better.

But fortunately all over the world people are taking action in hundreds of different ways to reduce their greenhouse gas emissions. Individuals are businessmen, politicians, civil servants, shopkeepers, and all sorts of professions and craftsmen. All of us are also consumers, making decisions like whether we really need another plastic bag from the supermarket or a four-wheel-drive all-terrain vehicle to take our children one mile down a city street to school.

Every day it gets easier to make the right decisions because information about new products for households and better support for technologies that avoid fossil fuel use are being made available. The level of information and education about the issues and benefits of tackling climate change have also improved dramatically. At city level, forward-looking mayors and city governments are transforming the quality of life for their citizens by providing better heating and lighting and reducing pollution and traffic, while at the same time helping the planet.

In writing this chapter I am referring to a booklet called the Little Green Book of Big Green Ideas, produced by Friends of the Earth, and to a compilation of 100 ways of saving the environment produced by Vanity Fair magazine. A few years ago it would have been unthinkable for these organisations to be on the same wavelength. The two publications include advice about a range of disparate ideas to help the environment, not just to combat climate change, but how to avoid using dangerous chemicals and eating endangered fish. Both are excellent and if half the population had a shot at only half the ideas then the world would be dramatically changed for the better. Other organisations like Greenpeace have long believed that just telling people what the problems are, apart from being depressing, leaves them feeling powerless. That is why the organisation had a solar demonstration plant mounted on the back of a lorry which toured Europe, and set up a wind power company so that the public can invest in and buy green power.

One of the problems of tackling climate change is that new technologies need mass uptake to achieve economies of scale. Computers and mobile phones are two examples of how this

Above: As well as the largest population in the world China also has the most cyclists, with a combined fleet of more than 500 million bikes. Total bike production is about 120 million units a year, many of which are exported. This bike park is in the capital, Beijing, where cycling is still the cheapest and healthiest form of transport. Although most people aspire to own a car, bikes are often as fast as cars on the clogged roads in the rush hour.

Above: German policemen watch members of the environmental group Robin Wood unfolding a banner at Berlin's Brandenburg Gate at the end of March 2006. The group was protesting against the use of fossil fuel and atomic power ahead of a German government energy summit and want the coalition of parties to phase out both as soon as possible. The banner reads: "Back to the soot", "Coal kills climate" and "The future is renewable".

happens when big business gets behind selling them. The price of renewable energy will fall dramatically when home energy kits are produced like fridges or washing machines, and that is beginning to happen.

This is because in Europe and over much of the world there are other factors apart from people power that are also pushing along the changes, particularly in dealing with climate change. There is increasing comprehension among governments that fossil fuel prices are going up and will continue to do so because supplies are not meeting demand. The only solution, for all countries, if their economies are not going to be damaged, is to reduce fossil fuel use, and the only way to do that is to develop alternatives.

For individuals, apart from a strong desire to reduce the dangers of global warming both for ourselves and future generations, there are other incentives. For the first time the cost of energy is beginning to make people think twice about the sort of car they buy and how much they need to heat and cool their homes. The fact that for the first time the average heating bills in Britain have topped £1,000 made headlines in 2006. Sadly the fact that this was more to do with the average Briton's profligate use of energy rather than the steep rise in gas prices got very little mention in the articles that followed.

But it remains true that rising energy prices are the greatest driver for change. It has been said many times, but it is worth repeating, that doing something about climate change is a massive boost to the economy, rather than the opposite, as President Bush and the fossil fuel lobby claimed. This boost applies to individual households as well as nations. It is bizarre that governments, charged with looking after the public welfare, acknowledge this, but so far

seem incapable of putting policies in place to take advantage of it. Britain's failure for 20 years to grasp the potential of wind power is an example. The Blair government, despite its rhetoric, continued to rig the market in favour of big generators rather than encourage much more climate-friendly micro-generation. Fortunately individuals seem to have a better grasp of economics than finance ministers. Pressure for change was building from below and legislation to promote fair play for small scale renewables was going through parliament in the summer of 2006 with all party support.

With 34 governments under pressure to meet their Kyoto targets, there has been a great deal of regulation on industry to reduce emissions. But one of the largest potential areas of savings, the use of power in homes, has hardly been addressed in most countries.

Germany is a shining example of what can be done. In 1998 Germany began a 100,000 solar roof programme, which gave people 10-year loans at reduced interest rates to buy photovoltaic systems. In five years the target had been reached and Germany had a competitive solar power industry. As well as boosting solar power, the German government at the same time embarked on encouraging wind power and overtook the rest of the world in installed capacity. More recently, as reported in the last chapter, the government decided to bring all its older properties up to modern energy efficiency standards at the rate of 5% a year. In 20 years all the older housing stock, mostly in the former East Germany, will be up to modern standards. It is a classic example of how jobs can be created where they are most needed, the housing stock improved and fossil fuel imports reduced all at the same time. Reducing Germany's greenhouse gas emissions

Top: The German chancellor, Angela Merkel, stands between the labour minister Franz Müntefering (right) and Brandenburg's state premier Matthias Platzeck as they pose for a picture during a cabinet meeting in the eastern German village of Genshagen on January 9th, 2006. They and other politicians in Germany's coalition government, seen behind, had just clashed over the size of a major new jobs and growth initiative and the future of nuclear power on the eve of a special two-day cabinet meeting, and wanted to show that despite this the government was united.

Above: In 2006 factories in Germany produced roughly 66% more solar cells than in the previous year, spurred on by the reforms made by the German chancellor. Here a man installs solar panels on a house, one of the many embracing the widespread domestic uptake of solar power.

and avoiding the need to replace the country's ageing nuclear reactors are two other advantages. Germany has the good fortune to have one of the most environmentally aware populations in the world. The result is that individual voters are helping to drive the changes by buying into the new technologies in large numbers.

The Japanese government, which has no fossil fuels of its own, and following the first oil price shock of 1973 became the most energy efficient industrialised country in the world, has also been a world leader in the new technologies. Japan began a solar roof programme before Germany and as a result became not only the world leader in solar energy production, but also the largest manufacturer and greatest innovator. In the early years of this century the sales of solar cells rose by more than 50% a year — and it is an industry that will outlast oil. Japan plans to get 10% of its electricity from solar cells by 2030.

While some technologies require government incentives to create a market volume that makes them economic, others are already economic but underused. An example is solar water heating, which in some countries, like Turkey, has long been a standard way of providing hot water for family homes and hotels. These solar systems do not produce electricity but simply heat water on a roof top and circulate it into a tank. A two square metre panel on a family home can reduce annual water heating bills by 70% and in any sunny countries, the Mediterranean area being an example, solar water heaters make economic sense without any government intervention.

However, some populations still need a push to change their habits. Spain, which has the vision to create jobs by becoming a leading

manufacturer of solar tube systems for heating water, made them compulsory from 2005 on all new buildings, domestic and commercial, substantially reducing the nation's energy bills. China, which already has the largest installed capacity of solar water heaters in the world, plans to quadruple it in 10 years.

But while governments have an obvious role to play in regulation and tax incentives to stimulate a mass market and to provide business confidence to create a manufacturing base in all these technologies, there needs to be co-operation and creativity across society.

Builders have been slow to adapt to new building techniques unless driven by regulation, frustrating a new breed of architects who design houses that need little or no energy to make them comfortable. While it is possible to retrofit solar and water heating panels and mini-wind turbines, nearly all the technologies are cheapest to install in new build. Increasingly popular on office blocks and private homes are green roofs. These provide roof gardens and insulation. In some more ambitious schemes water from showers and washing machines can be recycled via the roof garden, both growing plants and purifying the water at the same time. A reed bed on the roof can be a wildlife haven and a water cleaning system at the same time.

These and other technologies, mentioned below, need to be built into both commercial and domestic homes in the planning stage to gain most benefit. Many local authorities are already realising that builders will not do this on their own because many of the innovations, while cutting the running costs of buildings, add to the initial construction costs. To force builders to employ forward-looking architects, who have cut their environmental teeth on individual

Right: Iceland is a mass of volcanic activity and the hot rocks are very close to the surface, giving its people the chance to produce power from this abundant energy source with no damaging emissions to the environment. This geothermal power plant is the first in the world to combine the production of hot water with the production of cheap electricity. It is owned jointly by the Icelandic treasury and the local people and provides power and heating to local houses and industry. It is also a tourist attraction with a visitors' centre and bathing in the hot water from the plant which is full of minerals. It is said to provide a cure for skin diseases.

designs, planning authorities are imposing conditions on the mass market in order to cut emissions. An increasing trend in many European countries is to refuse planning permission for new housing schemes that do not have at least 20% renewable energy technologies built into their designs. Of course it is also in the interests of house purchasers to inquire what steps have been taken to protect them from excess use of fossil fuels, which are set to continue rising in price. All the evidence from across the world is that environmental benefits built into house design are a selling point. European law now requires an energy efficiency assessment of every house sold and details of energy bills.

In most countries there are demonstration projects, and frequently whole suburbs in developed countries designed to save energy and reduce environmental impacts. Many have become tourist attractions. China, which has a mass urbanisation programme, has gone one step further. It is preparing to build a whole city as an eco-project. With millions of Chinese now aspiring to city life in a western style they are planning to use renewable energy and combined heat and power plants for an eco-city in Dongtan, near Shanghai. The Chinese government is hoping that the city will be without the water shortages, air pollution and waste problems of many of its other rapidly developing urban centres.

Where it seems impossible to change your way of life or in the short term the way you run your business it is possible to buy carbon offsets. This is another growing business across the world, and potentially a great benefit both for green technologies and for forests.

The idea is simple. You calculate how much carbon dioxide is released into the atmosphere as a result of your activities, then you pay a company to invest in projects which saves the same amount of carbon elsewhere. The London newspaper the Guardian, which flies reporters round the globe for a variety of work, from covering the Iraq war to fashion shoots, is calculating and offsetting the carbon produced in those flights by investing in clean technology. An example would be funding a small scale hydroelectric power station in India, which would provide carbon-free power and light and stop the use of inefficient, expensive and polluting kerosene heaters. An individual feeling bad about the damage to the planet because of a long distance air flight for a holiday could pay a small sum to offset the carbon dioxide released.

The first and most popular form of carbon offsets in both Europe and America is to plant trees. A project could be in the home country, producing new native woodlands, or replacing felled tropical woodland in some far off continent. In every case the point is that while the tree grows it takes carbon dioxide out of the atmosphere and fixes it in the wood. In some cases just paying to preserve forest that would otherwise be felled is regarded as an option. Other land use projects which prevent carbon being released from the soil can also qualify. The problem with these schemes is that the science of exactly how much carbon is saved is not clear cut, and unscrupulous operators have made spurious claims, and even sold off newly planted woodland. Woods can also be cut down, catch fire or be "harvested", releasing the carbon back into the atmosphere.

As previously mentioned, renewable energy projects are a good option, with solar, wind, and biomass as well as small scale hydro all qualifying. These have the added benefit of allowing communities to develop without going

Left: A Chinese worker transports scrap metal to a recycling factory in Foshan in Guangdong province in southern China in February 2006. He is one of an army of individuals and companies attempting to slake China's insatiable thirst for raw materials.

Top: Another worker attempts to capitalise on another of China's problems, this time waste. This load of plastic bottles and containers balanced on a single hand cart are for recycling in Haikou, the capital of the southern province of Hainan. With a fifth of the world's population, the amount of waste in China has become an enormous problem.

About half of waste plastic is left uncollected or dumped in an uncontrolled manner on land, in rivers or in the sea, according to a 2003 collaborative study by the country's Institute for Environmental Studies and the Chinese Academy of International Trade and Economic Cooperation.

down the carbon-intensive route, and avoiding many of the health risks of indoor pollution. More suitable for larger companies rather than individuals are carbon capture projects. Greenhouse gases like hydrofluorocarbons and nitrous oxides which are released in manufacturing processes can be captured and destroyed. Methane, "harvested" from landfills, as described previously, can be burned to produce electricity and most ambitious of all, carbon dioxide from power stations is captured and pumped underground.

Another variation of that is to buy tonnes of carbon in the European carbon trading schemes, and simply take them off the market. This effectively reduces the pool of carbon available, puts the price up and makes it more difficult for polluting industries to buy their way out of taking action to curb their own emissions.

Perhaps the least tangible but most effective of all the options is energy efficiency. Measures could be as simple as providing low energy light bulbs or insulating material for roofs for a poor community that would otherwise not be able to afford them. This is as easily and cheaply done in your home country as abroad.

The problem with all these schemes is that for the layman they are hard to evaluate and verify. It is therefore essential that an independent assessment is built into all these schemes to prevent inappropriate investments and rogue traders.

This book was conceived by dakini books and myself not only to change people's attitude towards climate change and make people realise how urgent the problem is, but also to describe what can be done about it and set a good example. Lucky Dissanayake, the

publisher, sought expert assistance to help counter-balance the carbon emitted during the book's development and production. Sensible measures include using energy efficient appliances, turning computers off when not in use, and buying renewable energy from green suppliers. However, while such steps reduce emissions and should be adopted and encouraged, no business can entirely cease using energy or emitting carbon. The office lights often needed to be on late to ensure deadlines were met, and transporting the book inevitably requires fuel. Even more damaging to the environment, some air travel for research and development of the project was unavoidable. Despite the book being printed on paper from certified sustainable forests, so that the trees used for pulp were replanted, the most efficient paper manufacturer and printer is likely to emit more than a book's weight in carbon emissions.

Acknowledging this, dakini took advice. The idea was to cancel out the cost in carbon emissions of producing the book. The publisher has made a donation on the world emission trading market equal to the carbon released. An investment will then be made in the type of projects we have already described, which are certified under the Kyoto Protocol's clean development mechanism. This means the production of this book will help to pay for reduction of emissions in developing countries. The system allows businesses and individuals to make sure their activities have a neutral carbon impact on the environment, and provides a solution for those wanting to do their part in making rapid progress in tackling man-made climate change.

This is a practice that will become more widespread, although whether it will ever achieve the aims of a long-running and laudable

Right: To people from a developed country this is an appalling sight but for the people who work at this plastic recycling depot in Bangladesh this is a job which provides them with money to live. There are thousands of families for whom earnings from recycling by hand is a welcome alternative to no income at all. The millions of bottles have to be retrieved from rubbish, sorted into colours, cut up, and washed in the river Buriganga before being recycled.

"The ideal that human beings must strive for is not to conquer nature... but to live as a part of nature, in accord with its rules."

Kisho Kurokawa, architect, in thesis
The Symbiosis of Man and Nature, 1987.

campaign by Aubrey Meyer, of the Global Commons Institute, is debatable. His idea is to allow everyone in the world an individual carbon budget. The starting point is that the average American emits 20 tonnes of carbon dioxide each year, the average European 11 tonnes, a Chinese 2.4 tonnes and an Indian just over 1 tonne. Africans produce on average even less.

Aubrey's idea is a carbon allocation for the entire world, on the basis of a cut in man-made emissions of 60%. This total is then divided between countries based on the number of citizens that live in it. Over this century each country should reach its allocation. This would allow poor countries to increase their carbon output for the time being as they develop while the already industrialised countries adopt new clean technologies to reduce their carbon footprint. He calls it contraction and convergence. The idea has been widely praised as a possible way forward in international negotiations but so far, for many countries, mostly the profligate emitters, it seems too tall an order.

However, across most of the developed world there are ever-increasing numbers of books, guides and advice centres on how individuals can make a contribution to reducing their environmental imprint on the planet. Some measures are so simple, and obvious, that it seems superfluous to mention them. But the fact is that most of us simply do not think about it, still less make a conscious effort to do anything about it. So below are a few of the most straightforward.

The first and easiest way of saving energy and money is by turning off lights and machinery that are not needed. In many countries, including the UK, where the electricity market has been reorganised, it is possible to buy "green" electricity. This means stipulating that all the electricity you buy comes from renewable resources, for example wind, solar, hydroelectric or tidal.

Beyond that the bold next step of producing your own energy requires some thought. In each place and in each country conditions are different. It might be not be sunny or windy enough in some places for renewable energy to be economic, but there is bound to be at least one technology, which has already been developed, that is available where you are.

Small scale hydro-plants for example can be run for the benefit of one house or an entire village. In many countries grants are available for individual households or community ventures. Free advice is often provided by local authorities, and governments, to make sure schemes that seem attractive are viable. But these are for more ambitious projects. To start with many of the actions that individuals can take need no approval from anyone apart from yourself; they involve very little effort, and often tiny financial outlay, or none at all. Below are some of the many ideas aimed at individuals making a contribution to saving the planet.

The importance of cutting the consumption of energy in the home is a difficult area for politicians, and they have largely left the public in limbo, urging them to be responsible, but not helping much with what this means in practice. This is changing with the advent of home energy audits. This involves inviting a qualified person into your home whose job it is to assess your energy rating. Systems vary but one has a score between 1 and 120. The higher the score the better the quality of energy efficiency — 1 being the equivalent of standing in an open field. The average British house comes in at 46 on this

Left: Architects everywhere are trying to make their buildings practical and environmentally friendly, and still be beautiful. The goal of sustainable development is often hard to achieve, especially when building airports. This is part of what Malaysia says is the world's largest airport, with five runways, 60km (35 miles) south of Kuala Lumpur. It connects to the capital by train and there are internal driverless trains to take passengers between terminals. To offset the environmental costs a tropical rainforest was transplanted from elsewhere to surround the airport, in addition to vegetation being planted in the central garden space of the satellite area of the development.

Above: An artist's impression of Zhengzhou, a new Chinese city with a population of 2 million, which has ambitious plans to turn itself into a green city. It is the capital of Henan province, which has a population of 100 million, and is located in the middle reaches of the Yellow river. It is at the heart of the culture of central China. In a new plan for the city announced in 2001 the architects appointed by the city, Kisho Kurokawa, have proposed turning the banks of all 34 rivers in the region into parks to form a network of ecosystems to preserve biodiversity. They have also proposed using an existing fish-breeding pond to form China's largest artificial lake with a surface area of 800 hectares (2,000 acres) as a centrepiece of the city. The authorities hope to integrate river boats, road vehicles and a rapid transit system into new public transport for the city. They hope to complete the new city by 2015.

scale, but some northern European countries rate as high as 80. Other systems use an A to G scale, the same as the energy rating system for some electrical appliances, A being best. Since every house is different, and many were built before the age of building regulations, remedies to improve your score, on any of these scales, come in all shapes and sizes. The auditor's job is to suggest the simplest most cost-effective way of improving the score, and so reduce energy bills. It should also tell you, if you spend say £100, how long it will take you to get the money back in reduced bills. Many of the remedies you can do yourself, for example putting in loft insulation, or fitting draught excluders in ill-fitting window frames. More complex operations like cavity wall insulation clearly need specialist help, although almost without exception, where it is possible to put in cavity wall insulation where there is none, it is very cost effective.

Windows are a different matter. They too have energy ratings but the return on capital in terms of lower heating bills can take a lifetime. On the other hand if you are replacing windows anyway because they need it, getting the best insulated type makes a noticeable difference to comfort.

Saving costs in generating heat, whatever the condition of the fabric of the house, is clearly important. For those who use gas, a condensing boiler can cut bills by 20% to 30% overnight. These use the re-circulation of combustion gases within the boiler to extract the heat for hot water before the exhaust fumes are vented to the atmosphere. They are now a legal requirement in Europe, when you come to replace your boiler, and increase the efficiency of domestic gas burning for heating from 60% to nearly 90%. These energy audits are becoming compulsory in many countries when

people sell their homes, in order that the buyer can discover what his average household bills are likely to be. The object is to give the purchaser the opportunity to ask the seller to fix the problems before he buys or request a reduction in price so that he can make the improvements himself when he moves into the new home. But that bargaining just deals with the fabric of the home — roofs, walls, windows and floors — and fixtures like the boilers. The audit should also cover appliances that people might take with them. Perhaps the simplest and most overlooked of all the energy uses in the home is lighting.

Lighting accounts for 30% to 50% of the average building's energy use. Long-life light bulbs use one fifth of the energy to provide the same amount of light without wasting heat. They keep rooms much cooler in summer too. Fifteen years ago when I first replaced all the bulbs in my house with the long-life variety my electricity bill was halved overnight. The electricity board sent an investigator round to see if I was bypassing the meter but he departed as soon as he saw the long-life bulbs. The bulbs were guaranteed for five years but I did not have to change a light bulb for 12 years, by which time my investment had been repaid at least twice over. New types of long-life light bulbs are smaller, are graded into hard and soft lights and come in all shapes and sizes to fit all modern requirements. It is extraordinary that everyone does not use them. If you want to cut your bills further, turn off unnecessary lights, or if you are very lazy install sensors that turn the lights on when you come into the room, and off again when you are not there.

With electrical appliances of all kinds, particularly televisions, computers, music players of all sorts, and charging equipment for mobile phones and

Top and right: Swiss Re's London headquarters at 30 St Mary Axe, known to almost everyone as the gherkin, opened in May 2004 with the boast that it would use half the power of a normal skyscraper. It has a series of shafts that cool the building in summer and distribute passive solar heat in winter. It was designed by the architects Foster and

Partners to be London's first environmentally sustainable tall building. Among the building's most distinctive features are its windows, which open to allow natural ventilation to supplement the mechanical systems for a good part of the year. Whenever possible, recycled and recyclable materials have been specified throughout the building.

Swiss Re, the reinsurer, has done a lot to draw attention to the perils of climate change, and commissioned the building to demonstrate that distinguished buildings could also be environmentally sustainable.

other devices, turn them off when not in use. Many use almost as much power on standby as they do fully operational.

Most houses are full of electrical equipment, which is frequently either being added to or replaced. Throughout Europe and most developed countries on an increasing range of goods there is a legal obligation on manufacturers to tell customers how much energy the appliance they are buying uses. In the European system, as detailed above, A is best and G is bad. It has had an interesting effect on the manufacturers as now almost no appliances below a C rating are made any more. For the buyer it is best to go for as high a rating as possible to cut bills, but beware: even this can be misleading. What is also important is to consider the size of the fridge, freezer, washing machine or other piece of equipment. A large one of anything, whether it is A rated or not, is going to use a lot more power than a smaller machine. You should also consider what your requirements really are. A smaller model needs less space, is probably cheaper and saves money and energy, and might meet all your requirements.

All these small decisions about which sort of equipment to use can save you a lot of money, and the planet a lot of carbon dioxide emissions, without altering your lifestyle in the slightest — except perhaps saving you the effort of changing light bulbs.

But there are lots of other adjustments that can save more. Perhaps the simplest and most effective is turning down the thermostat on the central heating. Some houses are suffocatingly hot, and the occupants wear almost no clothes. In America, and increasingly in other countries, the reverse is true in the summer. People are

obliged to wear sweaters indoors because the air conditioning is so fierce. The cost of keeping houses too warm is high not just in energy but in health. For young people it is better for the heart to beat a little faster and burn off more energy in a slightly lower temperature.

There are lots of other simple methods of saving money with central heating, for example turning off radiators in rooms that are not occupied, or parts of the house that are not in use during the day. If heating controls are fitted in each room, as they are with most modern systems, temperatures can be set to suit the use the room is being put to. The heating in many buildings, both domestic and offices, can now be controlled remotely. For example, if home heating is turned down or off and you know you are going to arrive home in two hours it is possible to turn it on, or up to the required temperature, by accessing the controls through your mobile phone. There are dozens of other small tips that are both obvious and easy to forget — for example not filling the kettle too much when you boil water for tea or coffee.

Nearly all the suggestions so far come into the category of helping to save money and the planet without altering your lifestyle in the slightest, except perhaps to make yourself more comfortable. Other more ambitious efforts do involve making a bit of a statement as well as an additional outlay.

As noted earlier it is possible in many countries to buy green electricity through the grid, which is sometimes slightly more expensive. At least one method of generating your own electricity at home is now a feasible option anywhere in the world. If you live in a sunny country, Australia, or southern Europe, or almost anywhere in the tropics, producing electricity

Left: The Fibropower station, Suffolk, UK, designed by the architects Lifschutz Davidson Sandilands in the early 1990s to look as much like a grain silo as possible to fit into the farming landscape of Suffolk in England. The Fibropower generating plant was the world's first to be fuelled by poultry litter and now produces electricity for 12,500 local homes. Nothing goes to waste. At the end of the process the raw material has been turned into nitrate-free fertiliser, ready to be returned to the land.

direct from the sun with solar panels on the roof is becoming economically more viable with each year. Heating water this way is already cost-effective almost everywhere. Over much of Europe and America micro-wind turbines plugged into the home energy supply are also becoming a sensible option to get free electricity from the elements. There are already a number of products on the market and they are getting better all the time. Governments are slowly altering the regulations to make this a more viable option.

As already mentioned, those lucky enough to have access to a river or stream could try mini-hydropower. Since technologies like solar power have a 20-year guarantee with them, and will last for far longer, many people regard the investment in renewables as a sort of pension fund of free power. There are other ways of gaining energy from the environment like ground source heat pumps. This is a way of extracting the earth's natural heat by using a heat pump to raise the temperature to warm your home. Currently the payback period for heat pumps does not justify the capital cost unless you intend to stay somewhere for a long time, or they are being constructed new at the same time as your house. Having your own source of electricity or heating is, however, a good selling point for a property.

Outside the home there is also much that can be done. Transport is one of the great contributors to greenhouse gases. Again there are options which save money and do not change your lifestyle, and those which make a statement of intent. Perhaps the best example is the kind of vehicle you choose to drive. Almost all new cars, irrespective of engine size, easily travel up to 70 mph or 110 kilometres an hour, which, with the exception of Germany, is at or above the normal

top speed allowed by law. The popularity of 4x4 off-road vehicles to drop the kids at school and do the shopping is therefore just a very expensive, and in this context anti-social, fashion statement. The fact that they are often involved in accidents, particularly fatal ones of pedestrians, including school children, is causing a backlash in Europe. Many people are demanding that they should attract extra taxes to reflect the damage they do to people and the environment, and in some places such as Britain owners already face higher road taxes.

A more positive way for governments to encourage individuals is to give financial incentives to buy or use energy efficient cars. This is already happening in some cities where electric and gas-driven cars do not have to pay parking fees and have access to areas where traditional cars are excluded. Hybrid cars, which use a combination of petrol and electric motors, are under development by most manufacturers following the success of the Toyota Prius. These cars are silent for most city driving.

In America, following hurricane Katrina, the queue to buy the dual electric/petrol Prius was in sharp contrast to the inability of Ford and General Motors to offload their extremely low miles per gallon sports utility vehicles (SUVs). Consumers remain reluctant to buy SUVs despite desperate price-cutting measures, and both American giants are trying to modernise their models before they go under.

All sorts of new technologies, including cars driven by hydrogen-powered fuel cells and even compressed air, are under development. Ultimately the idea is to create a system that allows the car owner to gain tomorrow's "fuel" by plugging in his vehicle to an electric socket when he goes to bed at night. Whatever

Top: Honda is another Japanese vehicle manufacturer which realises the potential of hybrid technology. Already the hybrid system is in use on motorcycles and trains as well as cars. All the world's major manufacturers are now working on models of their own.

Above: Cars that run on hydrogen are often billed as a solution to the twin problems of pollution and climate change. When hydrogen is burned in fuel cells the only waste is pure water. This technology is no distant dream because over 50 million tonnes of hydrogen are produced and consumed every year and fuel cells work efficiently.

But as with other new ideas the problem is switching to a different world where motorists can feel there is a practical alternative to buying petrol and diesel vehicles and have confidence that when they drive into a filling station there will be the alternative fuels they need. Shell has a global hydrogen business and is committed to moving

production and sales from an industrial setting into the lives of ordinary people. Widespread use of hydrogen could also help address concerns about energy security. This is Shell's hydrogen fuel retail site at Benning Road in the US capital Washington.

"It [the Toyota Prius] has become an automotive landmark: a car for the future, designed for a world of scarce oil and surplus greenhouse gases."

Alex Taylor, Fortune,
February 2006.

Above: The 2004 Toyota Prius made its world premiere at the New York Auto Show on April 16th, 2003. The Prius is a "full hybrid system" car that can operate in either petrol or electric modes as well as one in which both the petrol engine and the electric motor are in operation at the same time. The battery that provides the electric drive is charged up every time the motorist brakes. After hurricane Katrina struck in 2005, fuel prices rose and an awareness of climate change grew in the US, dealers had to start a waiting list to buy the Prius because demand was so great. The car does as much as 50 miles per gallon in city traffic.

technology the driver has chosen the car will be recharged with electricity, restocked with hydrogen or filled with compressed air in time for the next morning's drive to work. While the world waits for this new generation of vehicles to arrive it is still easy to save money and reduce your current emissions by making the next family car as fuel efficient as possible. If bio-diesel or ethanol are available, even as part of a fuel mixture, use them. It is surprising how good it makes you feel running your car on old chip fat, or plants, and knowing that you are not damaging the planet in the process. It is, of course, often less stressful to leave the car at home and travel by public transport where possible. It is also healthier and cheaper to walk or ride a bicycle, particularly for short journeys — but for some that is still too revolutionary an idea for the 21st century.

Lots of other personal choices in our daily lives make a difference, for example what we eat. In the last 20 years it has become possible to pop down to the local supermarket and buy virtually any meat, fish, fruit or vegetable in any season. Much of it is flown across the world to meet the whims of shoppers for out of season straw-berries or peas. These so-called food miles are adding greatly to environmental costs, partic-ularly air flights. They have increased 23% in 30 years and supermarkets are largely to blame.

This could have disastrous consequences for us all. As well as the increase in greenhouse gases, the local farmers, who cannot sell their products because of the imports, are put out of business. This is usually not straight competition. Cheap labour or near slave labour in the developing world, or (the reverse way round) massive farm subsidies in Europe and America distort world prices. The result of all that has been the loss of the tradition of local produce for local needs.

At last, however, the consumer has revolted, or at least enough of them have done so to create an attractive alternative to supermarket shopping. In many parts of Europe the strong tradition of local produce has survived, particularly in France, which has remained more self-sufficient than any other European country. Elsewhere a new desire for local produce has grown into the farmers' market movement where farmers set up their own stalls in local towns to sell direct to the public. Another big development in Britain is the "box scheme". This is where locally produced food, often organic fruit and vegetables, are delivered to your door each week. Again it is direct selling of local produce in its season. This might seem a departure from the topic of climate change, but as the price of fossil fuels rises, and the land for growing food is lost to sea level rise, drought and deserts, local food supplies will become more important. Best to keep the option open — apart that is from having healthier food to eat and supporting your local community.

One of the great problems in dealing with man-made greenhouse gas emissions is the issue of flying. All over the world air travel is growing, and with it, emissions high in the atmosphere where scientists believe the pollutants magnify the problem. Some believe that the quantity of aircraft contrails or vapour trails in the sky actually mask the effect of global warming, as we saw when air travel stopped in the US following the destruction of the twin towers in New York. This might be an argument for allowing the increase in air flights to continue while we try to solve some of the environmental problems on the ground, but it is not one I would support. Currently limiting the growth of cheap air travel seems near impossible in any case. This is because the United States refuses to accept any tax on aviation fuel to reflect the

Left, top: Depletion of fish stocks across the world has led to some chefs and supermarkets insisting on buying only from sustainable resources. Increasingly fishing organisations are realising that the past hunt for short term profits has been killing their industry. This view of fishermen at sea with seagulls overhead is fast becoming a memory in some areas where fishing is banned because of over-fishing or stocks are so depleted it is no longer worth putting out to sea. The Sea Fish Industry Authority now works across all sectors of the UK seafood industry to promote good quality, sustainable seafood to try and keep the industry alive.

Left, bottom: In the UK and across Europe there has been an explosion of inter-est in organic farming and markets where farmers sell local produce direct to people who live in towns nearby. FARMA, the National Farmers' Retail and Market is a co-operative of people selling on a local scale. FARMA works throughout the UK and is the largest organi-sation of its type in the world, representing direct sales to customers through farm shops, pick-your-own schemes, farmers' markets and home delivery.

environmental damage air travel causes. Optimists believe that the problem will begin to take care of itself. As there is no alternative to fossil fuels for powering aircraft the cost of escalating oil prices will immediately feed through to ticket prices. Air travel and foreign holidays in distant places will begin to slip back into being an occasional luxury rather than the cheapest option for weekend breaks and second home owners. People will holiday within driving or train distance of their homes.

In this chapter are just a few ideas of what individuals can do. Most good environmental ideas in any case also help towards reducing emissions of greenhouse gases. For example recycling, composting and not using the plastic bags offered automatically by supermarkets all have a climate change element as well as saving resources. One area, which we have not touched upon, is voter power. Many of the technologies mentioned above are held back because politicians have failed to change the tax system or create the changes in regulation to favour a cleaner way of life. Countries that have had the courage to ban or tax plastic bags have made a substantial environmental gain, but most governments have given in to pressure from supermarkets to continue the practice of giving bags away for the convenience of customers. The long term detriment of the environment continues to be ignored. Voters need to fight back.

The market in Britain has long been rigged in favour of large power generators which feed electricity into the grid and against those who install their own home systems and might have a surplus. But a combination of businessmen, realising the potential of these new technologies, and voter pressure from those wanting to buy them, has forced the government to act and ministers have promised to deal with the regulatory problems. In countries where there is already a good return on extra electricity generated and fed to other neighbouring consumers via the grid, the uptake from householders for these promising technologies is far greater.

There are dozens of other examples where enlightened governments, urged on by consumers and voters, can change habits and markets. In doing so they can literally change the world, the technologies that power it, and probably save the planet (or at least most of it) from impending disaster. It is a daunting challenge, but we can all help to do it.

Right, top: Children are the ones who will suffer most if we continue to alter the atmosphere and damage the natural environment. Even the youngest of students are concerned for the future. Here on June 6th, 2006 in Panama City students are marking World Environment Day. The banner reads, "Conserve our environment".

Right, bottom: Turtles are one of the oldest creatures on the planet but in many places are endangered because of interference with the beaches where they bury their eggs to hatch. To help them survive in the 21st century turtles are protected in many countries until they are large enough to survive in the wild. Here Colombian

Wayuu Indian students release turtles at Colombia's Camarones beach on World Environment Day in June 2006.

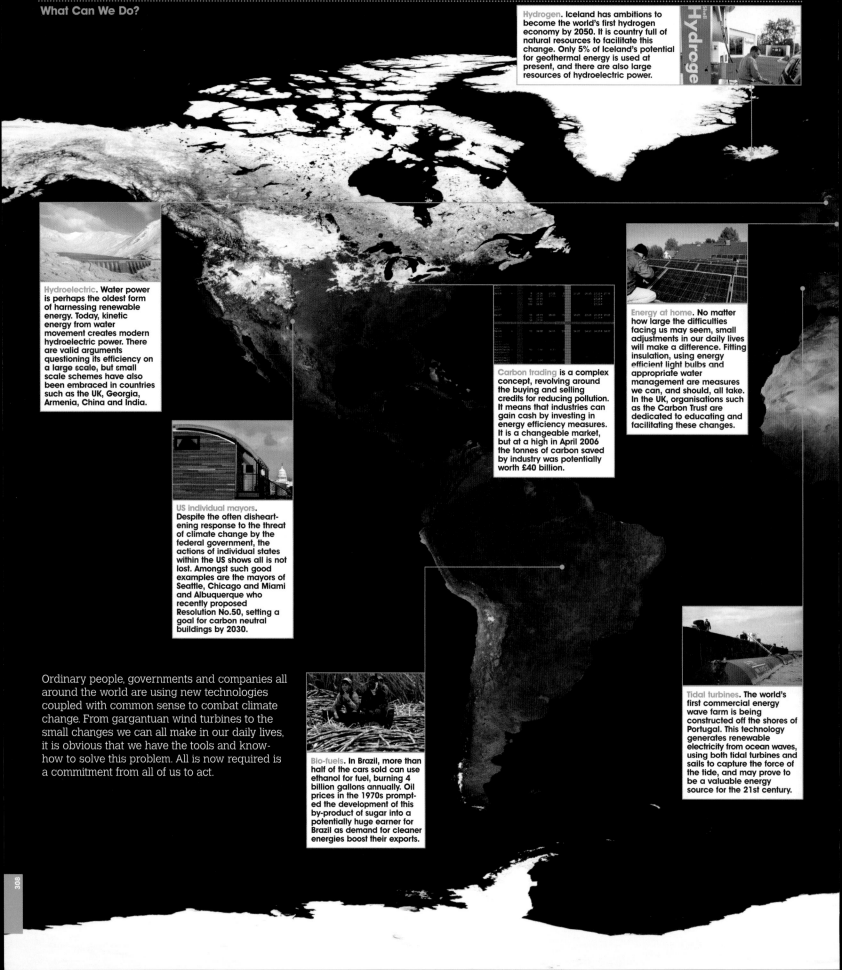

Hydrogen. Iceland has ambitions to become the world's first hydrogen economy by 2050. It is country full of natural resources to facilitate this change. Only 5% of Iceland's potential for geothermal energy is used at present, and there are also large resources of hydroelectric power.

Hydroelectric. Water power is perhaps the oldest form of harnessing renewable energy. Today, kinetic energy from water movement creates modern hydroelectric power. There are valid arguments questioning its efficiency on a large scale, but small scale schemes have also been embraced in countries such as the UK, Georgia, Armenia, China and India.

US individual mayors. Despite the often disheartening response to the threat of climate change by the federal government, the actions of individual states within the US shows all is not lost. Amongst such good examples are the mayors of Seattle, Chicago and Miami and Albuquerque who recently proposed Resolution No.50, setting a goal for carbon neutral buildings by 2030.

Carbon trading is a complex concept, revolving around the buying and selling credits for reducing pollution. It means that industries can gain cash by investing in energy efficiency measures. It is a changeable market, but at a high in April 2006 the tonnes of carbon saved by industry was potentially worth £40 billion.

Energy at home. No matter how large the difficulties facing us may seem, small adjustments in our daily lives will make a difference. Fitting insulation, using energy efficient light bulbs and appropriate water management are measures we can, and should, all take. In the UK, organisations such as the Carbon Trust are dedicated to educating and facilitating these changes.

Ordinary people, governments and companies all around the world are using new technologies coupled with common sense to combat climate change. From gargantuan wind turbines to the small changes we can all make in our daily lives, it is obvious that we have the tools and know-how to solve this problem. All is now required is a commitment from all of us to act.

Bio-fuels. In Brazil, more than half of the cars sold can use ethanol for fuel, burning 4 billion gallons annually. Oil prices in the 1970s prompted the development of this by-product of sugar into a potentially huge earner for Brazil as demand for cleaner energies boost their exports.

Tidal turbines. The world's first commercial energy wave farm is being constructed off the shores of Portugal. This technology generates renewable electricity from ocean waves, using both tidal turbines and sails to capture the force of the tide, and may prove to be a valuable energy source for the 21st century.

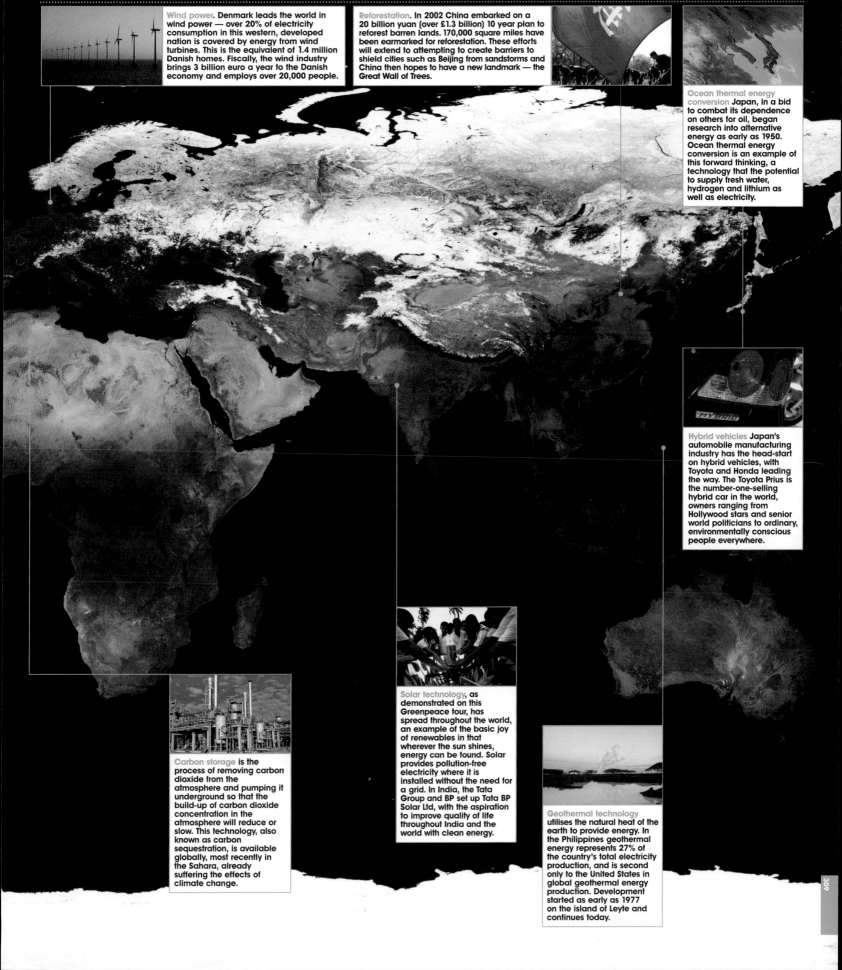

Wind power. Denmark leads the world in wind power — over 20% of electricity consumption in this western, developed nation is covered by energy from wind turbines. This is the equivalent of 1.4 million Danish homes. Fiscally, the wind industry brings 3 billion euro a year to the Danish economy and employs over 20,000 people.

Reforestation. In 2002 China embarked on a 20 billion yuan (over £1.3 billion) 10 year plan to reforest barren lands. 170,000 square miles have been earmarked for reforestation. These efforts will extend to attempting to create barriers to shield cities such as Beijing from sandstorms and China then hopes to have a new landmark — the Great Wall of Trees.

Ocean thermal energy conversion Japan, in a bid to combat its dependence on others for oil, began research into alternative energy as early as 1950. Ocean thermal energy conversion is an example of this forward thinking, a technology that the potential to supply fresh water, hydrogen and lithium as well as electricity.

Hybrid vehicles Japan's automobile manufacturing industry has the head-start on hybrid vehicles, with Toyota and Honda leading the way. The Toyota Prius is the number-one-selling hybrid car in the world, owners ranging from Hollywood stars and senior world politicians to ordinary, environmentally conscious people everywhere.

Carbon storage is the process of removing carbon dioxide from the atmosphere and pumping it underground so that the build-up of carbon dioxide concentration in the atmosphere will reduce or slow. This technology, also known as carbon sequestration, is available globally, most recently in the Sahara, already suffering the effects of climate change.

Solar technology, as demonstrated on this Greenpeace tour, has spread throughout the world, an example of the basic joy of renewables in that wherever the sun shines, energy can be found. Solar provides pollution-free electricity where it is installed without the need for a grid. In India, the Tata Group and BP set up Tata BP Solar Ltd, with the aspiration to improve quality of life throughout India and the world with clean energy.

Geothermal technology utilises the natural heat of the earth to provide energy. In the Philippines geothermal energy represents 27% of the country's total electricity production, and is second only to the United States in global geothermal energy production. Development started as early as 1977 on the island of Leyte and continues today.

Index

Credits

About the Publisher

At dakini books we pride ourselves on the quality, originality and perennial nature of our books. An independent publisher based in London, dakini produces only a select number of titles each year, our guiding principle being the demand for the subject matter.

We believe that the tide of public opinion is focused on the most crucial issue of our times, the threat of climate change. 'Global Warning: The Last Chance for Change' represents not only our own commitment to the environment, but our commitment to move and inspire the public at large to protect the planet we all live in.

www.dakinibooks.com

Thanks

Many thanks to the following organisations and photographers, who kindly donated the images featured in this book:

NASA/ESA/Greenpeace/WWF/Amjad Abdulla/George Fischer/ Brian Knutsen/Andy Harvey/Alastair Dobson/Rob Webster/ John N. Newby/Soh Koon Chng/Mario Farinato/His Excellency Mr. Maumoon Abdul Gayoom, President of the Republic of Maldives/Coral Cay Conservation Trust/ Kenya Tourist Board/Scottish Power/Paul West/Transport for London/ European Commission/Chevron/BP plc/Oddgeir Karlsson/ Kisho Kurokawa/Foster and Partners/ Lifschutz Davidson Sandilands/Honda/Shell/Seafish/FARMA

Photo Credits

Since I started researching and writing this book, twelve months ago, the amount of carbon dioxide in the atmosphere has gone up by four parts per million. Time is running out.

Paul Brown,
July 24th, 2006.